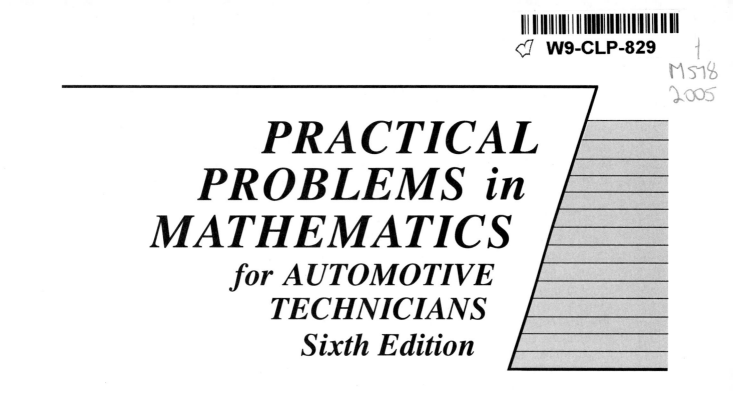

PRACTICAL PROBLEMS in MATHEMATICS

for AUTOMOTIVE TECHNICIANS

Sixth Edition

Delmar's *PRACTICAL PROBLEMS in MATHEMATICS* Series

PRACTICAL PROBLEMS in MATHEMATICS
for AUTOMOTIVE TECHNICIANS
Sixth Edition

Todd Sformo

Larry Sformo

GEORGE MOORE

THOMSON

DELMAR LEARNING ™

Australia   Canada   Mexico   Singapore   Spain   United Kingdom   United States

THOMSON

✦

DELMAR LEARNING

Practical Problems in Mathematics for Automotive Technicians, Sixth Edition

Todd Sformo, Larry Sformo, and George Moore

Vice President, Technology and Trades SBU:
Alar Elken

Editorial Director:
Sandy Clark

Senior Acquisitions Editor:
James Gish

Development:
Marissa Maiella

Marketing Director:
Dave Garza

Channel Manager:
Bill Lawrensen

Marketing Coordinator:
Mark Pierro

Production Director:
Mary Ellen Black

Production Manager:
Andrew Crouth

Production Editor:
Dawn Jacobson

Library of Congress Cataloging-in-Publication Data

Moore, George (George D.)
    Practical problems in mathematics for automotive technicians / George Moore.— 6th ed. / rev. by Larry Sformo and Todd Sformo.
        p. cm. — (Delmar's practical problems in mathematics series)
    ISBN 1-4018-3999-1
    1. Automobiles—Maintenance and repair—Mathematics. I. Sformo, Larry. II. Sformo, Todd. III. Title. IV. Series.
    TL154.M578 2004
    513'.024'6292—dc22

                                    2004011550

NOTICE TO THE READER

Contents

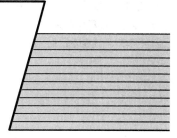

Preface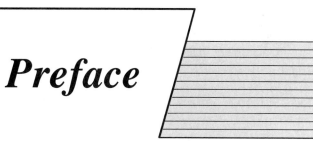

The automotive industry is well into the computer age. The dramatic changes of the last few years will seem insignificant when compared to those that are about to take place. The automotive technician, who will be called upon to service this new generation of automobiles, will have to be familiar with the automotive, scientific, and mathematical aspects of each system in a vehicle.

Practical Problems in Mathematics for Automotive Technicians, sixth edition, is a text–workbook that incorporates mathematical problems. The problems cover situations that automotive technicians encounter in routine service work, as well as situations encountered by specialists. By solving the problems, the automotive, scientific, and mathematical aspects are all strengthened, thus providing a solid foundation for a career as an automotive technician. Since many automotive students would like to own their own automotive service business, problems that deal with the operation of a business are used throughout the book.

The sixth edition of *Practical Problems in Mathematics for Automotive Technicians* has been revised to include estimating, more detailed explanations on how to work the problems, and more than 125 additional problems. Parts prices, interest rates, and labor costs are in line with today's economy.

The Instructor's Guide provides solutions and answers to all of the problems found in the text. Other instructional aids that will be helpful to the teacher are also included in the Instructor's Guide.

ABOUT THE AUTHORS

George Moore is an experienced automotive technician and taught in all areas of automotive technology at Aims Community College in Greeley, Colorado, for twenty years. An expert in the field, Mr. Moore has earned the title of Certified Master Automotive Technician from the National Institute for Automotive Service Excellence. He holds BA and MA degrees from Colorado State University. Now retired, he devotes his time to automotive consultant service, writing, and teaching emissions classes for the Colorado Department of Health.

Larry Sformo is a retired teacher and principal who has taught mathematics and science from K–12 for more than thirty years. He earned his BA and MA degrees and Certificate for Advanced Study from the State University of New York, Brockport. He is an adjunct instructor in Education at Medaille College in Buffalo, New York.

Todd Sformo is currently a PhD student at the University of Alaska, Fairbanks. He earned his BA from John Carroll University, a MA from the State University of New York at Buffalo, and an MFA and MS from the University of Alaska at Fairbanks.

ACKNOWLEDGMENTS

The authors would like to thank the following instructors for their reviews and helpful suggestions.

    John Burks, American River College
    Mike McCullough, East Mississippi Community College
    Walt Gunster, Tennessee Tech Center at Murfreesboro

Thanks also to Joanne Kirkpatrick Price for use of materials in her book *Basic Math Concepts: For Water and Wastewater Plant Operators,* Technomic Publishing Co. Inc., 1991.

TIPS FOR ESTIMATION

Being able to estimate is a very useful tool that we use every day in our personal and professional lives. For example, before going to eat in a restaurant, we estimate how much money we will take to pay for the meal. If we are going to a fast food restaurant, we will probably need less than $10. On the other hand, if we go to a fancy restaurant, we will probably need $30 or more.

Another example of estimating can be found in painting a car. We need to estimate how much paint to buy and how much money will be needed to purchase the paint. A pint of paint would not be enough for an entire car. We probably would need a gallon at least. A good enamel car paint can be very expensive. A gallon of house paint may be about $25, while the enamel car paint may be closer to $100. Before going to purchase the paint, we had better make sure that we have about $150 for the paint and other supplies.

As we can see, estimation of needs and resources can be very helpful in meeting our daily needs. It can save time and money. Estimation can be used in all math operations like addition, subtraction, multiplication, division, fractions, and decimals. We will see how the use of estimation can save us time in planning projects, save us time in doing homework, and save us money by not buying more than we need.

One method of estimating requires that we know how to "round" a number. Rounding of numbers gives us ballpark figures to work with. Oftentimes we only need to be close in our estimation. So rather than waste a lot of time figuring out the exact number, we can use estimation to make an educated guess.

How precise we need to be in estimating will decide which place values we will work with. Knowledge of place value is important in estimating. If we need to estimate how many sheet metal screws we are going to need for a job, then we can estimate to the nearest hundreds place. On the other hand, if we need to estimate how many tires to stock in the shop, then we probably want to estimate to the tens place. If we were to estimate tire stock to the hundreds place, we would have too many tires in the store, and it would cost us more money to purchase these extra tires. Estimating to the tens place will save us both valuable shop space and money on inventory.

Whenever we estimate a number, we always round it to a place value that will give us the best-estimated value. Let us quickly review the place values names.

Place Value	thousands	hundreds	tens	**ones**

So the number 46 is equal to 4 tens and 6 ones, or

tens	ones
4	6

The number 546 is equal to

hundreds	tens	ones
5	4	6

There are two simple rules to remember whenever we round a number:

First, we have to determine the place value of our estimate. Will it be the tens place? Hundreds place? Thousands place or higher? We call this the *chosen number* (CN).

For instance, if we want to round 238 to the nearest tens place, we see that the number 3 is in the tens place, so the 3 is the *chosen number* or CN.

Second, the CN will either *hold* (remain the same number), or it will be *pushed* up one number higher and become a 4. How do we know whether the chosen number remains the same (holds) or is pushed up? We look at the number to the right of the chosen number.

- If the number to the right of the CN is 0, 1, 2, 3, or 4, then the CN remains the same, and **all numbers after it become zero.** We call a number from this group a *hold number* (HN).

- If the number to the right of the CN is 5, 6, 7, 8, or 9, then the CN increases by one number, **and all numbers after it become zero.** We call a number that increases the CN by one a *push number* (PN).

Here is an example. We are asked to round 238 to the tens place.

First, find the CN. In this case, it is 3.

	CN	
2	3	8

Second, is the number to the right of the CN a *hold number* or a *push number*?

Remember: The number to the right of the CN is a hold number if it is 0, 1, 2, 3, or 4. The number to the right of the CN is a push number if it is 5, 6, 7, 8, or 9.

	CN	HOLD or PUSH?
2	3	8

In our example, the number to the right of the CN is an 8, so 8 is a push number (PN) which pushes the CN up one. The 3 becomes a 4, and all other numbers to the right of it become zero: 238 rounded to the tens place becomes 240.

Next, let us look at a number to the right of the CN that is a hold number (HN). In this case, we are asked to round 672 to the tens place.

First, the CN is a 7.

Second, the number to the right of the CN is a 2, which falls within the group of numbers called the hold numbers. Since the 2 is a hold number, the CN remains a 7 and all numbers to the right become zeros. We, therefore, have the number 672 rounded to the tens place as 670.

Try this example: Round 861 to the tens place.

First, what is the CN? _____

Second, the number to the right of the CN is one. Is this number a push number or a hold number? _____

Since it is a _____ number, the CN becomes _____. All numbers after the CN become zero.

The final answer is _____.

Try another example: Round 5,852 to the hundreds place.

First, what is the CN? In this case, the CN is 8.

Second, is the number to the right a hold or a push number? In this case, it is 5.

Since 5 is in the push group, the CN increases by one number, **and all numbers after it change to zero.** Our estimate of 5,852 would be 5,900.

Round the following to the nearest tens place.

1. 68	2. 31	3. 576
4. 12	5. 877	6. 125

Round the following to the nearest hundreds place.

1. 5,639	2. 1,283	3. 5,682
4. 9,765	5. 2,409	6. 12,863

Round the following to the nearest thousands place.

1. 45,673	2. 98,476	3. 75,537
4. 577,729	5. 34,321	6. 123,865

QUIZ:

1. What does CN stand for?

2. List the numbers that *hold* the CN.

3. List the numbers that *push* up the CN.

4. Bonus question: Round 4,567,358 to the ten-thousands place.

5. Double bonus question: We just bought a wrecked classic car. It has to be towed on a trailer to the garage for a total renovation. The car is located 634 miles away. We usually get about 14 miles per gallon when we tow the trailer. Question: About how many gallons of gas will it take for the trip? (**Hint:** Round 634 to the nearest hundreds place, and round 14 to the tens place.)

6. Triple bonus question: After we put the classic car on the trailer and towed it home, we found that we got 11 miles per gallon. Estimate how many gallons of gas we used for the trip home. Using that figure, estimate how many gallons we used for the round-trip. If gas is $1.73 per gallon, how much would you estimate that it cost us for the round-trip? (**Hint:** Round $1.73 to the nearest dollar.)

Whole Numbers

UNIT 1 ADDITION OF WHOLE NUMBERS

BASIC PRINCIPLES OF ADDITION OF WHOLE NUMBERS

Whole numbers refer to complete units with no fractional parts. Place the numbers in a column and add the units, starting from the right-hand column as in the examples below.

Example: Add the following: 12 + 1 + 25

Place Value	tens	ones
	1	2
		1
+	2	5
	3	8

Answer: 38

Example: Add the following: 12 + 1 + 25 + 9

Place Value	tens	ones
	1	2
		1
	2	5
+		9
	4	7

Answer: 47

To get 47, you need to "carry" numbers that add up to more than nine in each column. First, add the numbers in the column on the far right, which totals 17. Carry the 1 to the next column on the left, and add it to this column for the final answer.

Example: Add the following: 12 + 1 + 25 + 9 + 345

Place Value	hundreds	tens	ones
		1	2
			1
		2	5
			9
+	3	4	5
	3	9	2

Answer: 392

To get 392, you need to carry numbers that add up to more than nine in each column, just as you did in the previous example. First, add the numbers in the column on the far right, which totals 22. Carry the 2 to the next column on the left, and add it to this column to get 9 in the tens place. Finally, add numbers in the hundreds place, if necessary.

Suggestion: If you are using lined paper, turn the paper so that the lines run up and down instead of left to right. Now you have columns. Place one number in one column as shown in the three examples and add. This will keep the numbers in their proper place values.

ESTIMATING IN ADDITION

Reminder: Being able to estimate can save time and money on jobs. It can also help with the math in this book. If we estimate answers before trying to find exact answers, we can compare our estimates with the exact answers. If the estimated answer and the exact answer are not close, we will know that something is wrong.

Suppose we are shopping for auto parts and need to know how much money to take with us. Estimate how much the following purchase will cost. Round to the tens place.

Muffler	$58.00
Tailpipe	$15.00
Clamps	$13.00

We estimate the cost of the purchase as follows:

	COST				
CN	HOLD or PUSH?			ESTIMATE	
5	8	$\Rightarrow$		6	0
1	5			2	0
1	3		+	1	0
			$	9	0

The actual purchase amount is $86.00, before taxes—which is close to our estimate of $90.

Estimating allows us to figure out approximately what the answer should be. We can then check our answers to make sure that they are reasonable by comparing the estimate and the exact answer. In the auto parts example, if we had made a mistake when finding the exact answer and had come up with $63 instead of $86, then our estimate would have told us we needed to recheck our addition. This technique becomes more important when we have to decide whether we need to multiply or divide numbers.

Here is another example:

	EXACT					
CN	HOLD or PUSH?			ESTIMATE		
7	3				7	0
2	5	$\Rightarrow$			3	0
+ 1	6			+	2	0
1 1	4			1	2	0

PRACTICAL PROBLEMS

Estimate and give exact answers to problems 1–3 and 7–9. For the rest of the problems, give exact answers only.

1. 2 + 16 + 43 =   _____

2. 9 + 16 + 23 =   _____

3.	337	4.	2	5.	333	6.	13
	26		37		717		691
	17		18		85		55
	+ 3		+ 559		+ 9		+ 14

7. In wiring a trailer, the following lengths of wire are used: 9 feet, 3 feet, 8 feet, and 10 feet. What is the total length of wire used?

_____

Many different types of *bolts* are used in an automobile.

HEX HEAD
BOLT

HEX HEAD
SHOULDER BOLT

8. In repairing a car, 6 hex head bolts, 8 oil-pan bolts, 3 fender bolts, and 2 differential-cover bolts are needed. How many bolts are used in all?

_____

9. In taking inventory, the following number of hose clamps are checked: 100 radiator-hose clamps, 72 heater-hose clamps, and 15 gas-line-hose clamps. What is the total number of clamps on inventory?

_____

10. On five different trips, an automobile is driven 33 miles, 27 miles, 19 miles, 42 miles, and 24 miles. What is the total mileage?

_____

11. In one week a driver uses 10 gallons, 6 gallons, 2 gallons, 9 gallons, 8 gallons, 3 gallons, and 12 gallons of gasoline. What is the total number of gallons used?

_____

12. A mechanic removes 20 bolts from the cylinder heads, 16 bolts from the oil pan, and 7 bolts from the flywheel housing. What is the total number of bolts removed?

_____

13. Headlights draw 10 amperes of current, an ignition coil 4 amperes, taillights 2 amperes, and a car heater 8 amperes. What is the total amperage drawn from the battery?

_____

Bolts are fastened with *nuts.*
Nuts are made in various shapes and have different uses.

SQUARE OR CARRIAGE NUT   PAL NUT   SLOTTED OR CASTELLATED NUT   CAP OR FINISHING NUT   WING NUT

14. A parts clerk has 16 of the ⅜-inch SAE nuts in stock and orders 134 additional nuts. How many nuts will there be when the new order arrives? _____

15. A person drives a car 129 kilometers to one city, 134 kilometers to another, and 219 kilometers home. What is the total distance? _____

16. A mechanic needs the following lengths of ⅜-inch copper tubing: one piece, 15 inches; one piece, 8 inches; one piece, 10 inches; one piece, 11 inches; one piece, 16 inches; one piece, 9 inches; two pieces, 12 inches each. How many inches of tubing are needed for the job? _____

UNIT 2 SUBTRACTION OF WHOLE NUMBERS

BASIC PRINCIPLES OF SUBTRACTION OF WHOLE NUMBERS

Subtraction is the process of finding the difference between two numbers. Place the smaller number under the larger number, keeping the right-hand column even (or justified) as in the examples below.

Example: Subtract the following: 15 − 5

Place Value	tens	ones
	1	5
−		5
	1	0

Answer: 10

Start with the right-hand column and subtract the smaller number from the larger number. In this case, 5 − 5 = 0, and you can bring down the 1, since there is no other number in the tens column to subtract.

Example: Subtract the following: 25 − 3

Place Value	tens	ones
	2	5
−		3
	2	2

Answer: 22

Start with the right-hand column and subtract 3 from 5. In this case, 5 − 3 = 2, and you can bring down the 2, since there is no other number in the tens column to subtract.

Example: What happens if you had to subtract the following: 35 − 6?

Place Value	tens	ones
	3	5
−		6
	2	9

Answer: 29

In this case, you start with the right-hand column, but now you cannot subtract 6 from 5, so you must "borrow" one from the number to the left.

First, borrow 1 from the 3 in the tens place (3 − 1 = 2), and put the 1 you borrowed in front of the 5. This will give you 15 (see below).

Second, 15 − 6 = 9, and you carry the 2 down from the tens place to get 29.

Place Value	tens	ones
	3̸ 2	15
−		6
	2	9

Example: Subtract the following: 1,934 − 345

Place Value	thousands	hundreds	tens	ones
	1	9	3	4
−		3	4	5
	1	5	8	9

Answer: 1,589

To get 1,589, you follow the same procedure as in the last example. Since you cannot take 5 away from 4 (in the ones place), you borrow from the number to the left, which gives you 14 − 5 = 9 in the ones place.

Place Value	thousands	hundreds	tens	ones
	1	9	3̸ 2	14
−		3	4	5
				9

Next, in the tens place, you see that you cannot take 4 away from 2, so you must borrow from the number to its left, which is in the hundreds place. This means that 9 becomes 8 and the 2 in the tens place becomes 12, so 12 − 4 = 8.

Place Value	thousands	hundreds	tens	ones
	1	9̸ 8	3̸ 12	14
−		3	4	5
		5	8	9

Now you can subtract 3 from 8 in the hundreds place to get 5 and bring down the one in the thousands place to get the final answer of 1,589.

ESTIMATING IN SUBTRACTION

When shopping for auto parts, we need to know how much money to take with us. Estimate how much the following purchase will cost. Round to the tens place.

Muffler $58.00

Rebate $15.00

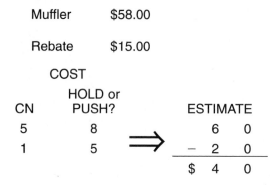

COST

CN	HOLD or PUSH?		ESTIMATE	
5	8		6	0
1	5		− 2	0
			$ 4	0

The actual amount is $43.00, which is close to our estimate of $40.00. Estimating allows us to figure out approximately what the answer should be, and then check our answers to make sure they are reasonable by comparing the estimate and the exact answer. If we had made a mistake when finding the exact answer, then we could recheck our subtraction. This technique becomes more important when we have to decide whether to multiply or divide numbers.

Here is another example:

EXACT

CN	HOLD or PUSH?		ESTIMATE	
7	3		7	0
− 2	5		− 3	0
4	8		4	0

PRACTICAL PROBLEMS:

Estimate and give exact answers to problems 1–3 and 9–11. For the rest of the problems, give exact answers only.

1.	553 − 89	2.	87 − 14	3.	349 − 327	4.	1347 − 250

5.	454 − 178	6.	351 − 269	7.	5,033 − 269	8.	125,360 − 53,451

9. If 39 quarts of oil are removed from a stock of 350 quarts, how many quarts are left?   _____

10. Driver number one drives 169 kilometers while driver number two drives 272 kilometers. How many kilometers farther does number two drive than number one?   _____

11. In a period of a month, one driver uses 93 gallons of gasoline while another driver uses 54 gallons. How many more gallons of gasoline does the first driver use than the second?   _____

Washers are placed under the head of a screw
or bolt to prevent looseness.

FLAT WASHER   STANDARD LOCKWASHER

EXTERNAL TOOTH   INTERNAL TOOTH
LOCKWASHER   LOCKWASHER

12. If 47 washers are taken from a box containing 126 washers, how many washers are left?   _____

13. A bill for repairs is $247. A discount of $24 is given. How much does the customer pay for the repairs?   _____

14. An automobile is driven 31,103 miles, of which 11,392 miles are on company business. How many miles are driven on personal business?   _____

Note: Use this diagram for problem 15.

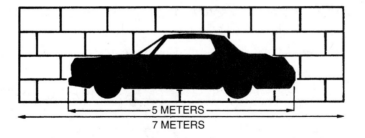

5 METERS

7 METERS

15.   A car measures 5 meters in length. The garage is 7 meters long inside. How much total clearance, in meters, is there when the car is parked in the garage?   _____

16.   An inventory sheet shows 165 pounds of body putty. In the first week 24 pounds are used; in the second week, 17 pounds are used. How many pounds are left?   _____

17.   A stock clerk fills orders for 6, 8, 4, 16, 12, and 24 spark plugs from a stock of 153 plugs. How many plugs are left in stock after filling the orders?   _____

18.   A 250-foot coil of ignition wire is taken out of stock. Lengths of 2 feet, 3 feet, 1 foot, 3 feet, and 2 feet are cut off. How many feet of wire are left in the coil?   _____

19.   A driver's odometer reads 17,565 before leaving on a trip. At the end of the trip, the odometer reads 17,823. How far has the driver driven?   _____

20.   A driver's starting odometer reads at 232,213. At the end of a haul, it reads 233,066. How far was the haul?   _____

UNIT 3 MULTIPLICATION OF WHOLE NUMBERS

BASIC PRINCIPLES OF MULTIPLICATION OF WHOLE NUMBERS

Multiplication is a simplified way of adding the same number many times. Instead of adding a column of numbers that are all the same, write the number and under it the number of times it is to be added, as in the examples below:

Examples:

Place Value	tens	ones
	1	0
×		4
4	0

Place Value	tens	ones
	1	4
×		2
2	8

Place Value	hundreds	tens	ones
		1	0
×		1	2
		2	0
	1	0	
	1	2	0

Place Value	ten-thousands	thousands	hundreds	tens	ones
			5	4	6
		×		8	3
		1	6	3	8
	4	3	6	8	
	4	5	3	1	8

For the last example, starting at the right side, multiply 3 times 6, which equals 18. Put down the 8 and carry the 1. Multiply 3 times 4, which equals 12, plus the 1 that was carried over, which equals 13. Put down the 3 and carry the 1. Multiply 3 times 5 and add the 1, which equals 16. Then multiply 546 by 8, but place your answer under the 8. After you have multiplied all numbers by the second number, add together for the final answer. The final answer is called the product.

PRACTICAL PROBLEMS

Estimate and give exact answers to problems 1–3 and 7–9. For the rest of the problems, give exact answers only.

1. 19
    × 9

2. 190
    × 28

3. 73
    × 34

4. 351
    ×   7

5. 643
    × 27

6. 8,811
    × 346

7. 234
    × 378

8. 10,093
    ×   18

9. The speed of a car is held constant at 54 miles per hour. How far does the car travel in 8 hours? _____

10. A car is traveling at 80 kilometers per hour. How many kilometers does the car cover in 7 hours? _____

11. A car moves 9 feet in one revolution of a wheel. How many feet does it move in 427 revolutions of the wheel? _____

12. The gasoline tank in a car holds 19 gallons. The car averages 26 miles to a gallon. How many miles does the car travel with one tank of gasoline? _____

13. A car uses gasoline at the rate of 1 gallon to every 27 miles. How many miles does the car travel on 199 gallons of gasoline? _____

14. A coil from an automobile has 57 layers of wire wound on it with 346 turns per layer. How many turns of wire are in the coil? _____

These *trim clips* are used on two different makes of automobiles.
They hold the upholstered door panel on the door.

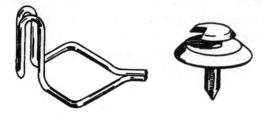

15. There are 9 trim clips used to hold the upholstered door panel on one door. How many trim clips are needed to hold the panels on all four doors of 14 cars? _____

16. A car travels at the rate of 84 kilometers per hour for 5 hours on Monday, 73 kilometers per hour for 6 hours on Tuesday, and 65 kilometers per hour for 9 hours on Wednesday. What is the total number of kilometers driven?   _____

A *rivet* is a pin with a head used to hold parts together. A driver or hammer-and-rivet set is used to form the head at the other end.

A *pop rivet* is used when the end of the rivet cannot be reached for flattening. The head is formed without the use of a driver or hammer-and-rivet set.

FLAT HEAD RIVET   POP RIVETS

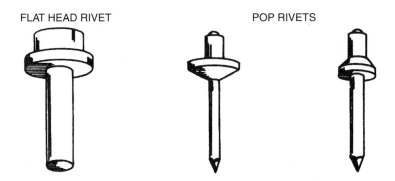

17. A body shop repairer estimates that 18 pop rivets are used on each body repair job. If 18 jobs are done per month, how many pop rivets must be ordered for 2 months' work?   _____

18. The lighting system of a car contains: 31 of the 2-candlepower bulbs, 2 of the 21-candlepower bulbs, 4 of the 6-candlepower bulbs, and 3 of the 15-candlepower bulbs. What is the total candlepower of the bulbs?   _____

UNIT 4 DIVISION OF WHOLE NUMBERS

BASIC PRINCIPLES OF DIVISION OF WHOLE NUMBERS

Division is a simplified method of subtracting one number from a larger number many times. The number to be divided (dividend) is placed inside the division frame. The divisor is to the left, and the answer (quotient) is above the frame, as in the example below:

Example:

$$\text{DIVISOR} \overline{)\, \substack{\text{QUOTIENT} \\ \text{DIVIDEND}}}$$

To divide 1035 by 23:

Example:

$$
\begin{array}{r}
4 \\
23\overline{)1035} \\
92
\end{array}
\qquad
\begin{array}{r}
4 \\
23\overline{)1035} \\
92 \\
\hline
11
\end{array}
\qquad
\begin{array}{r}
45 \\
23\overline{)1035} \\
92 \\
\hline
115 \\
115
\end{array}
$$

First divide 23 into the smallest number possible (103). Multiply 4 times 23, which equals 92. Subtract from 103, which equals 11. Bring down the 5. Divide 115 by 23, which equals 5. Multiply 5 times 23, which equals 115. There is no remainder.

PRACTICAL PROBLEMS

Estimate and give exact answers to problems 1–3 and 9–11. For the rest of the problems, give exact answers only.

1. $7\overline{)392}$
2. $23\overline{)92}$
3. $8\overline{)76}$
4. $3\overline{)159}$

5. $14\overline{)1190}$
6. $16\overline{)256}$
7. $49\overline{)735}$
8. $15\overline{)390}$

9. A garage owner buys 54 feet of heater hose. If each car uses 9 feet of hose, how many cars can be repaired? _____

10. An apprentice is paid $84.40 per day. How many days of work does it take to earn $754? _____

11. A woman drives her car 384 miles. If her car averages 32 miles per gallon, how many gallons of gasoline are used?   _____

12. A stock clerk has 256 spark plugs, and there are 8 plugs in a box. How many boxes of plugs are in stock?   _____

Electrical current jumps across the air gap in the spark plug
and a spark is made.

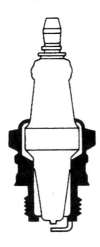

13. On a trip, a car uses 18 gallons of gasoline while traveling 378 miles. How many miles per gallon does the car average?   _____

14. A car weighs 7,216 kilograms. What is the average weight per wheel?   _____

Note: Use this diagram for problem 15.

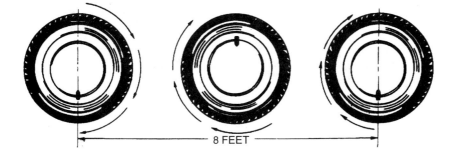

8 FEET

15. In one revolution of a wheel, a truck moves 8 feet. How many revolutions are required to move the truck 1 mile? (**Hint:** There are 5,280 feet in one mile.)   _____

16. Jim travels 7,021 miles, which is 7 times as far as Bill travels. How far does Bill travel?   _____

17. Two automobile mechanics, working together, spend a total of 240 hours on a job. If they work 8 hours a day, 5 days a week, how many weeks does the job take?   _____

18. The 24 valves on a 6-cylinder engine are reconditioned in 180 minutes. How much time does it take to recondition each valve?   _____

19. A garage owner uses 180 liters of motor oil to make oil changes in 36 automobiles. How many full liters (on the average) does each require?   _____

20. A driver travels 940 miles in three days. On Monday the car is driven for 5 hours, on Tuesday for 6 hours, and on Wednesday for 9 hours. What is the average speed for the trip?   _____

Decimal Fractions

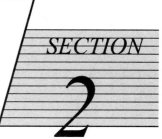

Unit 5 FRACTIONAL EQUIVALENTS

DECIMALS

Decimals are much easier to use than fractional measurements. Decimals can be given in the thousandths and ten-thousandths. Most automotive measurements are in the thousandths.

A decimal is a unit of measurement that means less than one. For instance, .005 means five-thousandths. Notice the "ths" at the end of the word. This is a signal that the measurement is less than one. In terms of fractions, which we will be discussed in the following units, we would write .005 as $\frac{5}{1000}$.

Recall the range of place values. (See the place value spectrum below.)

ten-thousands	thousands	hundreds	tens	ones	.	tenths	hundredths	thousandths	ten-thousandths
————	———	———	——	——	.	———	————	————	—————

To figure out how to read a decimal number, place the number in the blanks. *First,* make sure that you know where the decimal point is in your number. *Second,* place the numbers to the right of the decimal in the blanks to the right. *Finally,* read the numbers to the right of the decimal and then read the place value of the last number.

For instance, we already know how to say .005: five-thousandths. If we were to place the number in the place value spectrum, we would see the following:

ten-thousands	thousands	hundreds	tens	ones	.	tenths	hundredths	thousandths	ten-thousandths
————	———	———	——	——	.	0	0	5	—————

What if we have a complex decimal like the following: 6.0357. Follow the same procedure as above, but this time when reading the number, use the word *and* when you are at the decimal point. In this case, the number would fit on the spectrum as follows:

ten-thousands	thousands	hundreds	tens	ones	.	tenths	hundredths	thousandths	ten-thousandths
————	———	———	——	6	.	0	3	5	7

To change a fraction into a decimal, divide the top number (called the numerator) by the bottom number (called the denominator).

17

Example: $\dfrac{3}{4}$ is $4\overline{)3.00}$ with quotient $.75$

$$\begin{array}{r} 2\ 8 \\ \hline 2\ 0 \\ 2\ 0 \\ \hline \end{array}$$

$\dfrac{3}{16}$ is $16\overline{)3.0000}$ with quotient $.1875$

$$\begin{array}{r} 1\ 6 \\ \hline 1\ 40 \\ 1\ 28 \\ \hline 120 \\ 112 \\ \hline 80 \\ 80 \\ \hline \end{array}$$

PRACTICAL PROBLEMS

For these problems, use Table II in Section II of the Appendix.

1. Find the nearest fractional drill size to the following decimal measurements:

   a. 0.880 inch   _____   g. 0.234 inch   _____

   b. 0.555 inch   _____   h. 0.752 inch   _____

   c. 0.230 inch   _____   i. 0.029 inch   _____

   d. 0.618 inch   _____   j. 0.100 inch   _____

   e. 0.815 inch   _____   k. 0.680 inch   _____

   f. 0.439 inch   _____   l. 0.935 inch   _____

2. What measurement is $\frac{1}{16}$ inch larger than 0.750 inch?   _____

3. What measurement is $\frac{1}{16}$ inch larger than 0.250 inch?   _____

4. What measurement is $\frac{1}{8}$ inch larger than 0.625 inch?   _____

5. What measurement is $\frac{1}{8}$ inch larger than 0.375 inch?   _____

6. What size of a drill is $\frac{1}{16}$ inch larger than $\frac{1}{2}$ inch?   _____

7. What size of a drill is $\frac{1}{4}$ inch larger than 0.375 inch?   _____

8. What is the diameter of a $\frac{3}{8}$-inch hole drilled $\frac{3}{16}$ inch oversize?   _____

9. What is the diameter of a $\frac{3}{4}$-inch hole drilled $\frac{1}{16}$ inch oversize?   _____

10. What measurement is $\frac{1}{16}$ inch larger than $\frac{5}{8}$ inch?   _____

Unit 6 ADDITION OF DECIMAL FRACTIONS

BASIC PRINCIPLES OF ADDITION OF DECIMAL FRACTIONS

Place the numbers to be added under each other so that the decimal points line up. Add as you would with whole numbers, and drop the decimal point into the answer, directly in line with the other decimal points.

Recall the place values for numbers on both sides of a decimal point:

Place Value

thousands   hundreds   tens   **ones**   .   **tenths**   hundredths   thousandths

Notice that the decimal goes between the **ones** place and the **tenths** place.

Example: Add the following: 7 + 2.32 + 0.008

Step 1.   Rewrite the numbers so that the decimal points line up.

Step 2.   Add as many zeros as necessary *to the right* of the decimal point so that each number ends at the same place value.

Place Value	ones		tenths	hundredths	thousandths
	7	.	0	0	0
	2	.	3	2	0
+	0	.	0	0	8
	9	.	3	2	8

Answer: 9.328

Example: Add the following: 8.5 + 2 + 11 + 11.01

Step 1.   Rewrite so that the decimal points line up. Recall that a whole number like 2 is actually 2.0 and 11 is actually 11.0

Step 2.   Add as many zeros as needed to the right of the decimal point so that each number ends at the same place value.

Place Value	tens	ones		tenths	hundredths
		3	.	5	0
		2	.	0	0
	1	1	.	0	0
+	1	1	.	0	1
	2	7	.	5	1

Answer: 27.51

PRACTICAL PROBLEMS

1. What is the total length of the clutch plate pilot tool shown here?   _____

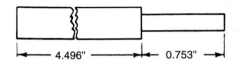

2. A valve will expand in length when it is heated. The length of the valve when cold is 4.8750 inches. It expands 0.0015 inch when heated. What is its length when hot?   _____

3. A 165 R14 radial tire costs $44.95, and the Federal Excise Tax (FET) is $2.09. What is the total cost of the tire?   _____

4. A customer at a gasoline station buys 1 quart of oil for $1.55, polish for $2.48, and a polish cloth for $0.79. What amount is paid to the attendant?   _____

5. A Nissan valve stem measures 102.35 mm. Maximum guide clearance is 0.045 mm. What is the largest permissible guide diameter in millimeters?   _____

6. What is the total amount of a bill itemized as follows: compressor seal, $8.75; refrigerant, $17.12; clutch brush set, $9.20; labor, $98.50; sales tax, $2.81?   _____

7. A starter ring gear with a 14.675-inch inside diameter will expand 0.075 inch in diameter when heated. What is the diameter when heated?   _____

8. A standard piston diameter is 72.967 mm. The largest oversize piston available is 1.55 mm larger than standard. What is the diameter of the oversize piston in millimeters?   _____

9. Five-thousandths of an inch are to be honed out of a bore that measures three and eight hundred seventy-five thousandths inches. What is the finished size stated numerically?

_____

Note: Use this diagram for problems 10 and 11.

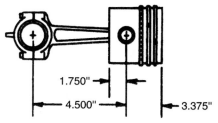

1.750"

4.500"

3.375"

10. What is the distance, in inches, from the center of the connecting-rod bearing to the top of this piston?

_____

11. What is the length, in inches, of this piston?

_____

Note: Use this diagram for problems 12 through 14.

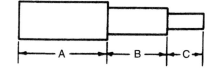

A   B   C

12. **A** = 3.125 inches, **B** = 2.015 inches, **C** = 1.500 inches. What is the total length in inches?

_____

13. If **A** = 3.250 inches, **B** = 2.375 inches, **C** = 1.250 inches, find the total length in inches.

_____

14. What is the total length of the pin, stated numerically?

_____

A = Three and seventy-five thousandths inches.

B = One and eight hundred seventy-five thousandths inches.

C = Eight hundred forty-five thousandths of an inch.

15. A chassis frame of a car is 0.250 inch thick, the spring hanger is 0.375 inch thick, and it takes 0.050 inch to head the rivet. What is the total length of a rivet needed to fasten these two pieces together? _____

   Note: The term *oversize,* as used in connection with parts having a circular section, such as pistons, piston pins, and cylinders, refers to an increase in the diameter of the part from the original or standard size. The term *clearance,* as used in connection with parts having a circular section, refers to the difference between the two diameters of the parts fitted together.

   Hint: When reboring a cylinder, metal is removed all around the cylinder circumference.

16. What size should the rebore be to straighten up this bore? _____

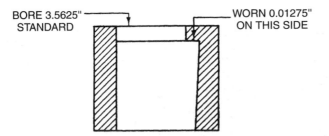

BORE 3.5625"
STANDARD
WORN 0.01275"
ON THIS SIDE

17. A piston pin wears a groove in a cylinder wall 0.012 inch deep on one side and 0.0075 inch on the other. The original size of the cylinder is 3.875 inches. What is the smallest oversize that this cylinder should be rebored? Note: Standard oversizes for reboring are: +0.010", +0.020", +0.030", +0.040", +0.050", +0.060". _____

18. A Jaguar 4.2-liter engine has a standard bore of 92.97 mm. The maximum wear in a cylinder is 0.238 mm over this size. What is the diameter, in millimeters, of the worn cylinder? _____

19. A master cylinder piston is 17.33 mm in diameter. The clearance is 0.13 mm. Find the diameter of the cylinder in millimeters. _____

20. The standard size of a certain piston pin is 0.925 inch; the oversize is 0.005 inch greater. What is the actual size? _____

21. A cylinder is rebored to 0.020 inch oversize. The standard size is 3.375 inches in diameter. What is the size of the rebored cylinder? _____

22. An aluminum piston expands 0.0015 inch when hot. What is its diameter after expansion if its standard diameter is 3.6875 inches? _____

23. A certain cylinder is 3.875 inches in diameter standard size. It is honed 0.005 inch oversize. What is the size of the honed cylinder? _____

24. A hole for a piston pin, originally 0.8175 inch in diameter, has worn 0.0012 inch. What size is the hole? _____

25. A 0.002-inch clearance is allowed on a 0.85-inch diameter truck spring bolt. What size is the bushing? _____

26. A standard cylinder bore is 3.4375 inches in diameter. If 0.0285 inch is removed by reboring, what is the diameter after reboring? _____

27. A cylinder bore measures three and six hundred eighty-seven thousandths inches and is worn six-thousandths of an inch. What is the original diameter of the bore, stated numerically? _____

28. A mechanic replaces a set of standard piston pins that measure 0.927 inch in diameter with a set 0.015 inch oversize. What do the new pins measure? _____

29. On a honing job, the original pistons, measuring 4.125 inches, are replaced with pistons 0.010 inch oversize. If a 0.002-inch clearance is allowed, what size is the honed cylinder? _____

30. An aluminum alloy piston expands 0.00075 inch in diameter when heated. The standard size is 3.9687 inches in diameter. What is its size after expansion? _____

31. A foreign car connecting rod has a pinhole diameter of 17.427 mm. The specifications call for 0.020 mm interference fit with the pin. What was the diameter of the pin in millimeters? _____

   Note: *Interference fit* means that the hole is smaller than the pin and must be pressed in with a press.

Unit 7 SUBTRACTION OF DECIMAL FRACTIONS

BASIC PRINCIPLES OF SUBTRACTION OF DECIMAL FRACTIONS

Place the smaller number under the larger number, keeping the decimal points in line. Subtract as you would with whole numbers. Drop the decimal point in the answer directly below the others.

Example: Subtract the following: 8 − 5.1

Step 1. Rewrite the numbers so that the decimal points line up.

Step 2. Put in as many zeros as necessary *to the right* of the decimal point so that each number ends at the same place value.

Place Value	ones		tenths
	8	.	0
−	5	.	1
	2	.	9

Answer: 2.9

Example: Subtract the following: 10.032 − 9.15

Step 1. Rewrite the numbers so that the decimal points line up.

Step 2. Put in as many zeros as necessary *to the right* of the decimal point so that each number ends at the same place value.

Place Value	tens	ones		tenths	hundredths	thousandths
	1	0	.	0	3	2
−		9	.	1	5	0
			.	8	8	2

Answer: .882, which can also be written as 0.882.

PRACTICAL PROBLEMS

1. How much must be cut from a shackle bushing measuring 2.375 inches in length to fit a spring 2.125 inches wide? _____

2. If 0.022 inch is removed from a pin 1.25 inches long, what length is left? _____

3. The standard width of a certain piston ring is 0.1875 inch. The micrometer reading is 0.184 inch. How much is it worn? _____

4. An aluminum piston from an antique car measures 3.875 inches when cold. When heated it measures 3.8815 inches. What is the amount of expansion? _____

5. Correct bearing preload on a differential is obtained when the distance between the bearing caps is 185.5 mm. When the mechanic measures the distance, it is 187.43 mm. How much more should the bearings be tightened? _____

6. Two pistons have diameters of 3.8125 inches and 3.875 inches. What is the difference in the diameters? _____

7. In the following illustration, dimension **B** is three hundred nine thousandths inch, and dimension **C** is four and seventy-five thousandths inches. Find dimension **A** numerically. _____

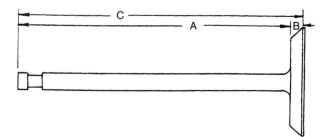

8. The difference in the sizes of a new 3.375-inch diameter piston and a worn one is 0.0185 inch. What is the size of the old piston? _____

9. The standard diameter of a brake drum is 200 mm. If 0.92 mm is removed from the diameter to correct scoring, what is the new diameter in millimeters? _____

10. A customer is charged $62.10 for labor and $24.83 for parts for a repair job on a car. What is the change from a $100 bill? _____

Note: Use this diagram for problems 11 and 12.

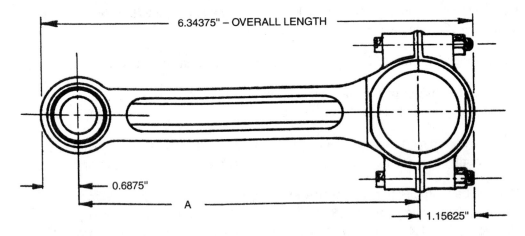

11.   What is the center-to-center distance, dimension **A**, between the holes?   _____

12.   If the overall length is 8.0000 inches, and the other two dimensions are 0.750 inch and 1.0050 inches, what is dimension **A**?   _____

13.   Find, in inches, length **X** on this skirt. The other dimensions are: _____

       B = 0.375"

       C = 0.090"

       D = 0.1875"

       E = 0.090"

       F = 0.125"

       G = 0.1875"

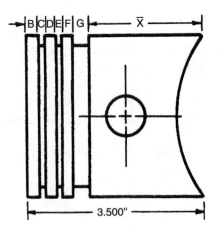

14. A starter commutator 2 inches in diameter is turned down on a lathe. The tool makes a cut 0.0625 inch deep. What is the finished diameter in inches? _____

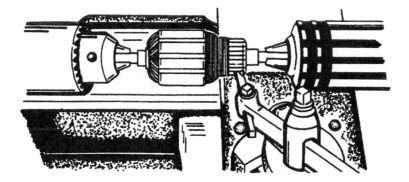

15. A certain foreign car manufacturer calls for tire pressure measurements in kilograms per square centimeter. The reading on one tire is 1.57 kg/cm^2. The specification calls for 1.72 kg/cm^2. How much air should be added to the tire? _____

16. A 0.0015-inch feeler gauge is placed between the cylinder wall and the piston. If the cylinder bore is 3.5 inches, what is the diameter of the piston in inches? _____

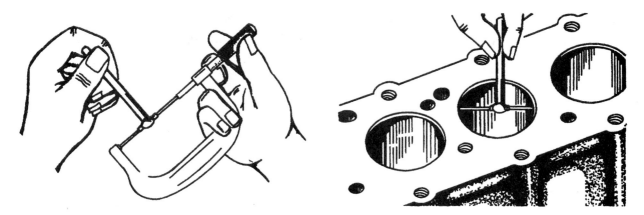

17. The standard size of a bore is 3.875 inches. It is honed to 3.877 inches. How many inches is the diameter increased? _____

18. A piston is to fit in a 3.875-inch bore with a clearance of 0.002 inch. What amount must be removed from a piston that is 3.883 inches in diameter? _____

19. A piston measures 3.8825 inches. It is to be placed in a bore measuring 3.875 inches with 0.0025-inch clearance. How much material must be removed to fit the piston in the cylinder? _____

20. A bore that is standard at 3.875 inches is bored to an oversize of 3.915 inches. How many thousandths of an inch must be removed?   _____

21. A cylinder bore that is out-of-round measures 2.8125 inches in one direction and 2.8160 inches in the other direction. How much is the cylinder out-of-round?   _____

22. A cylinder measures 0.020 inch over standard. The standard is 3.375 inches. What is the piston size for this cylinder, allowing 0.002-inch clearance?   _____

23. When ordering pistons, the jobber always wants to know the exact size of the renewed cylinders. A cylinder 4 inches in diameter was bored 0.030 inch oversize. What is the actual size of the piston that is received from the jobber if this product requires 0.002-inch clearance?   _____

24. A piston is to be finished to a specified size of 3.6835 inches. During the operation of finishing the piston, a test measurement is 3.685 inches. How much more stock must be removed?   _____

25. A 0.375-inch valve stem at point of greatest wear is 0.0175 inch undersize. What is its diameter at this point?   _____

26. The standard diameter of the spider in a universal joint is 14.72 mm. A worn joint measures 14.36 mm. How much, in millimeters, is the spider worn?   _____

Unit 8 MULTIPLICATION OF DECIMAL FRACTIONS

BASIC PRINCIPLES OF MULTIPLICATION OF DECIMAL FRACTIONS

Multiply decimal fractions in two steps:

Step 1. Multiply the numbers *as if* they were whole numbers. In other words, for step 1, pretend the decimal points do not exist.

Step 2. Count the number of places the decimals are from the last number in *both* numbers multiplied. In your answer, place a decimal point exactly the same number of places from the last number of the answer.

Example: **Step 1.**
$$\begin{array}{r} 6.37 \\ \times\ \ 2 \\ \hline 1274 \end{array}$$
Multiply as if there were no decimal point.

Step 2. Count the number of places the decimal is from the last number in 6.37. The decimal point is two places; therefore, place a decimal point in your answer two places from the last number. Your final answer should be 12.74.

Example: **Step 1.**
$$\begin{array}{r} 7.35 \\ \times\ \ 4.5 \\ \hline 33075 \end{array}$$
Multiply as if there were no decimal points.

Step 2. Count the total number of decimal places in *both* numbers. 7.35 has two places, and 4.5 has one place. This gives a total of three decimal places. In the final answer, place a decimal point three places from the last number. The answer should be 33.075.

PRACTICAL PROBLEMS

1.
$$\begin{array}{r} 3.5 \\ \times\ 2 \\ \hline \end{array}$$

2.
$$\begin{array}{r} 3.5 \\ \times 3.2 \\ \hline \end{array}$$

3.
$$\begin{array}{r} 79.01 \\ \times\ \ 2 \\ \hline \end{array}$$

4.
$$\begin{array}{r} 87.22 \\ \times\ .55 \\ \hline \end{array}$$

5.
$$\begin{array}{r} 31.273 \\ \times\ \ .02 \\ \hline \end{array}$$

Using Table II in Section II of the Appendix, express the following decimal fractions to the nearest indicated scale measurement.

6.   0.31 inch to nearest $\frac{1}{16}$ inch   _____

7.   0.809 inch to nearest $\frac{1}{16}$ inch   _____

8.   0.860 inch to nearest $\frac{1}{8}$ inch   _____

9.   0.781 inch to nearest $\frac{1}{32}$ inch   _____

10.   0.5155 inch to nearest $\frac{1}{64}$ inch   _____

11.   0.340 to nearest $\frac{1}{32}$ inch   _____

12.   0.185 to nearest $\frac{1}{16}$ inch   _____

13.   0.627 to nearest $\frac{1}{8}$ inch   _____

14.   0.980 to nearest $\frac{1}{64}$ inch   _____

15.   0.060 to nearest $\frac{1}{16}$ inch   _____

Note: Use this diagram for problems 16–21.

Pitch is the distance between the threads on a bolt or screw. In each turn,
a bolt or screw moves a distance to the pitch.

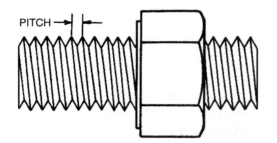

16.   The pitch of a thread is 0.125 inch. How far, in inches, is a nut moved in 6.5
turns?   _____

17.   How far, to the nearest $\frac{1}{16}$ inch, is the nut moved in 6.5 turns? (The pitch is
0.125 inch.)   _____

18.   If the pitch is 1.5 mm, how far, in millimeters, is a nut moved in 36 turns?   _____

19.   How far, in millimeters, is the nut moved in 16.93 turns? (The pitch is
1.5 mm.)   _____

20. A machine screw has 12 threads, and the pitch is 0.0625 inch. How long, in inches, is the screw under the thread? _____

21. How long, to the nearest $\frac{1}{64}$ inch, is the screw under the head? There are 12 threads with a pitch of 0.0625 inch. _____

22. A 6-cell storage battery on discharge gives a voltage reading of 1.85 volts per cell. What is the total voltage of the battery? _____

23. A 6-cell storage battery shows an average voltage of 2.11 volts per cell. What is the voltage of the battery? _____

24. An *ampere-hour* is the product of discharge rate (in amperes) and the time needed (in hours) to discharge a battery. The discharge rate for a battery is 11.25 amperes, and the time is 7.2 hours. What is the probable capacity of the battery in ampere-hours? _____

25. A gasket set for an air-conditioning compressor costs $23.61. The garage uses 3 sets per week. What does an 8-week supply cost? _____

26. Each brush set for a certain type of alternator costs $5.39. What is the total cost of the brush sets if 6 alternators are repaired? _____

27. A set of alternator brushes for an internal regulated alternator costs $12.10. What do a dozen sets of brushes cost? _____

28. A certain type of starter brush costs $7.39 for a set of 4 brushes. What will 8 sets of brushes cost? _____

29. A mechanic receives $17.20 per hour. How much is earned in 20 hours? _____

30. On a special deal, a parts store buys air-conditioning hose for $3.10 per foot. The hose must be bought in a job lot of 256$\frac{3}{4}$ feet. How much does this lot of hose cost to the nearest cent? _____

31. At a cost of $1.41 per foot, what does 9 feet 3 inches of air duct hose cost? Round the answer to the nearest cent. _____

32. A garage owner buys 3$\frac{1}{4}$ dozen cans of polish at $6.25 per can. What is the total cost? _____

33. A firm sold 114 automobiles last year at an average price of $16,759.61. What were the total receipts from the sale of cars? _____

34. The cost of gasoline is $1.639 per gallon. How much does 369 gallons cost
     to the nearest cent?   _____

35. A garage owner charges $65.00 per hour for labor. What is the charge for a
     job that takes 12½ hours?   _____

36. In a 6-cylinder car, the piston displacement of one cylinder is 38.667 cubic
     inches. What is the total piston displacement for the engine of this car?   _____

37. A man contracts for 36,550 gallons of gasoline at $1.375 per gallon. What is
     the total cost of the gasoline?   _____

38. A customer gives a mechanic a $50 bill in payment for ½ hour labor at $60.00
     per hour and a small replacement part at $8.76. How much change does the
     customer receive?   _____

39. A repair job takes 4⅛ days to complete. At $224.80 per day, what is the cost
     to the nearest cent?   _____

40. What is the weight of 24¾ gallons of gasoline if one gallon weighs 6.56
     pounds?   _____

41. A gallon of gasoline weighs 6.56 pounds. An empty gasoline truck weights
     4,500 pounds. What is the total weight of a truck carrying 450 gallons of gas?   _____

42. A gallon of grease weighs 7.44 pounds and costs $1.62 per ¼ pound. What
     is the cost, to the nearest cent, of a 53½-gallon barrel of grease?   _____

43. An air-conditioning clutch armature on a car is replaced. The cost of material
     is $75.90 and labor is ¾ hour at $60.00 per hour. What is the total cost for
     the repairs to the nearest cent?   _____

44. A dealer sells 5 quarts of oil at $3.00 a quart. The oil costs $1.10 a quart.
     What is the profit on the sale?   _____

45. What is the profit on 450 gallons of gasoline that costs $1.45 per gallon and
     sells for $1.59 per gallon?   _____

46. What is the cost of 720 gallons of gasoline at $1.62 per gallon?   _____

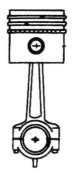

47. The weight of a certain piston and connecting rod assembly is 12.65 ounces. What is the weight, in ounces, of 8 assemblies?   _____

    Note: Use this diagram for problem 47.

48. In a tune-up on an 8-cylinder car, the materials used are: spark plugs, $2.95 each; points, $4.99; condenser, $2.43; and distributor rotor, $2.99. What is the total cost of the materials?   _____

49. A certain type of metal weighs 1.9 pounds per square foot. In building a truck body, 359 square feet of this metal are used. What is the total weight of the metal used?   _____

50. The cost of the metal used for a truck body is $1.16 per ½ pound. What is the total cost, to the nearest cent, if 663.8 pounds of metal are used?   _____

51. A mechanic overhauls a power steering pump using the following materials: seal kit, $16.25; 2 quarts of oil, $3.36 each; 2 hose clamps, $0.83 each; 2 hours labor, $45.00 per hour. What is the total cost of the job?   _____

52. A speedometer is repaired on a certain car. The materials used are: main frame assembly, $10.65; odometer, $13.00; cross shaft, $3.65; cable core, $4.56; 3 hours labor, $24.00 per hour. What is the total cost of the job?   _____

Unit 9 DIVISION OF DECIMAL FRACTIONS

BASIC PRINCIPLES OF DIVISION OF DECIMAL FRACTIONS

To divide decimals, make the divisor a whole number by moving the decimal point all the way to the right. Move the decimal point in the dividend to the right the same number of places. Keep the decimal point in the answer directly above the decimal point in the dividend.

Example:

```
                    13.0        (Quotient or Answer)
(Divisor)   6.4 ) 83.20        (Dividend)
                  64
                  19 2
                  19 2
```

PRACTICAL PROBLEMS

1. A family rents a truck for $86.40 per day. Based on an 8-hour day, what is the rental price per hour?   _____

2. A car travels 1,349.92 kilometers on 170 liters of gas. Find, to the nearest hundredth, the number of kilometers per liter.   _____

3. What is the average speed, in kilometers per hour, of a car that is driven 560 kilometers in 7.75 hours? Express the answer to the nearest hundredth.   _____

4. A truck driver charges $1.63 for each parcel that is delivered. In one day $208.64 is collected. How many parcels are delivered?   _____

5. A worker's paycheck for a 5-day workweek is $454.80. A workday is 8 hours. What is the worker's pay per hour?   _____

6. The cost of maintaining 12 small trucks for one year is $37,053.92. How much does it cost per week to maintain one truck?   _____

7. On a certain trip a car travels 372.6 miles and averages 38.5 miles per hour. How long does it take to make this trip, to the nearest tenth of an hour?   _____

8. It costs $140 to reline a set of brakes. The brakes last 38,000 miles. What is the cost per mile to the nearest tenth cent?   _____

9.   If 15 gallons of gasoline weigh 98.4 pounds, what does one gallon weigh?   _____

10.   A certain size tire costs $98.00. If the tire is used for 17,762 miles, what is the cost per mile? Express the answer to the nearest tenth cent.   _____

11.   A car uses 1 quart of oil each 425 miles. How many quarts of oil are needed on a trip of 2,525 miles? Round the answer to the nearest quart.   _____

12.   A car is driven 38,460 miles in one year at a cost of $7,192.02. What is the cost per mile?   _____

13.   A tire costs $78.70 and is used on a car for 14,235 miles. What is the cost per mile to the nearest tenth cent?   _____

14.   A used car dealer sells 14 automobiles for a total of $79,136. What is the average price of each car?   _____

15.   A set of spark plugs costing $15.68 is used for 11,512 miles. What is the cost per mile? Express the answer to the nearest tenth cent.   _____

Note: Use this diagram for problems 16 and 24.

Piston displacement (PD) is the volume (number of cubic units) displaced as the piston moves from bottom dead center (BDC) to top dead center (TDC).

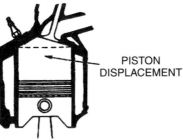

PISTON
DISPLACEMENT

Formula: *PD = Area of cylinder base × stroke length.*

16.   The piston displacement of a certain 4-cylinder engine is 140.3 cubic inches. What is the displacement of one cylinder?   _____

17.   The distance around the outside of a tire (circumference) is 92.250 inches. How many times will the tire rotate in 500 feet? Express the answer to the nearest hundredth.   _____

18. A pail of grease contains 42.5 pounds and costs $72.30. Find, to the nearest tenth cent, the cost per pound.   _____

19. Several pieces of heater hose totaling 42 feet cost $35.70. What is the cost per foot?   _____

20. The cost of 22 point sets is $108.90. What is the cost per set?   _____

21. A space on a steel frame is 14.885 inches long. This space is divided into 4 equal parts. How long is each space to the nearest thousandth inch?   _____

22. A driver is paid $56.90 for a 185.5-mile trip in an automobile. How much is the driver paid per mile? Express the answer to the nearest tenth cent.   _____

23. A car owner's garage bill comes to $285.80. One-half of the bill is paid in cash, and the remainder is paid in 5 equal payments. What is the amount of each equal payment?   _____

24. The total displacement in a 4-cylinder engine is 134.2 cubic inches. What is the displacement in each cylinder?   _____

25. An automobile travels at a rate of 46.5 miles per hour for 227 miles. How much time does the trip take, to the nearest hundredth hour?   _____

Unit 10 MICROMETER READING: APPLICATION OF DECIMAL FRACTIONS

BASIC PRINCIPLES OF MICROMETER READING

English micrometers divide the inch into one thousand parts. The barrel has a scale divided into 40 equal spaces of 0.025" each. The movable part, called the thimble, is divided into 25 equal parts of 0.001" each. One full rotation of the thimble moves it one full space on the barrel.

To read the English micrometer:

Step 1. Read the highest number showing on the barrel.

Step 2. Add 0.025" for each *full* vertical space showing.

Step 3. Add the number on the sleeve that lines up with the horizontal line on the barrel.

Example:

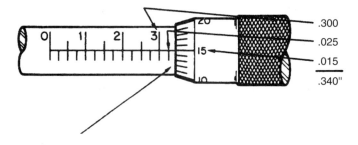

Caution: **Do not read this space as a full unit unless the zero is lined up with the horizontal line on the barrel.**

To read an inside micrometer, one additional step is required. The thimble only moves along the barrel ½ inch (0.500"). To make measurements between ½" and the next inch, a ½" spacing collar is installed on the inside micrometer.

Example:

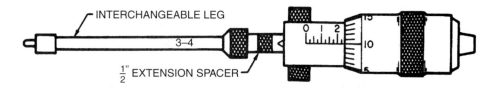

Metric micrometers are graduated with metric units. The horizontal line is divided into millimeters with graduations from 0 mm to 25 mm. Each millimeter is subdivided into 0.5 mm. The thimble is graduated in 50 divisions of 0.01 mm. One turn moves the thimble 0.5 mm. Two turns move it 1 mm. Metric micrometers are read similarly to English micrometers. The reading is obtained by adding the three decimal fraction values.

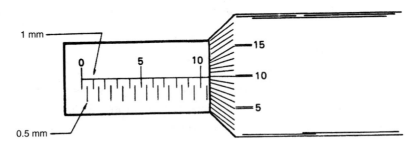

To read this metric micrometer setting:

Step 1.	Read the number of full divisions on the barrel.	10.00
Step 2.	Add the 0.5-mm divisions.	0.50
Step 3.	Add the thimble reading.	0.10
The reading is the sum.		10.60 mm

Apply these principles to the automotive field by solving the practical problems that follow.

PRACTICAL PROBLEMS

In problems 1–15, fill in the following tables showing how English micrometers are set to obtain the readings given.

Note: Use this table for problems 1 through 7.

1-INCH ENGLISH MICROMETER

Prob. No.	Reading in Inches	Number of numbered divisions on the barrel	Number of unnumbered divisions on the barrel	Number of thimble divisions
1.	0.875			
2.	0.3125			
3.	0.21875			
4.	0.893			
5.	0.666			
6.	0.008			
7.	0.077			

Note: Use this table for problems 8 through 15.

ENGLISH MICROMETER

Prob. No.	Reading in inches	Size of extension rod in inches	1/2-inch spacing collar		No. of numbered divisions	No. of unnumbered divisions	No. of thimble divisions
			Yes	No			
8.	2.625						
9.	6.1875						
10.	4.34375						
11.	3.750						
12.	2.555						
13.	4.067						
14.	5.899						
15.	3.999						

In problems 16–22, fill in the following table showing how metric micrometers are set to obtain the readings given.

Note: Use this table for problems 16 through 22.

Problem Number	Reading in millimeters	Number of numbered mm divisions of barrel	0.5-mm graduations beyond numbered mm divisions	Number of thimble divisions
16.	5.36 mm			
17.	21.21 mm			
18.	8.79 mm			
19.	19.67 mm			
20.	3.98 mm			
21.	25.40 mm			
22.	17.83 mm			

In problems 23 through 32, demonstrate your ability to read the English micrometers shown.

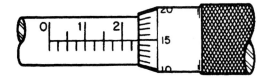

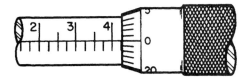

23. _____   24. _____

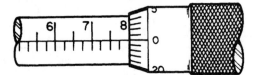

25. _____

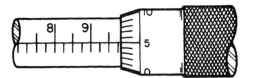

26. _____

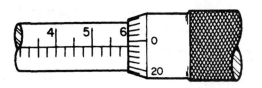

27. _____

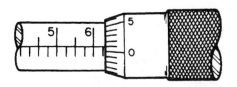

28. _____

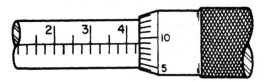

29. _____

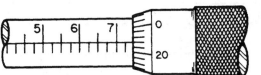

30. _____

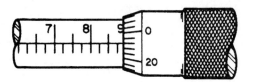

31. _____

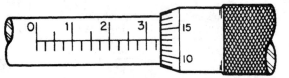

32. _____

In problems 33 through 37, demonstrate your ability to read the micrometers shown. Note that some micrometers have the ½" sleeve in place.

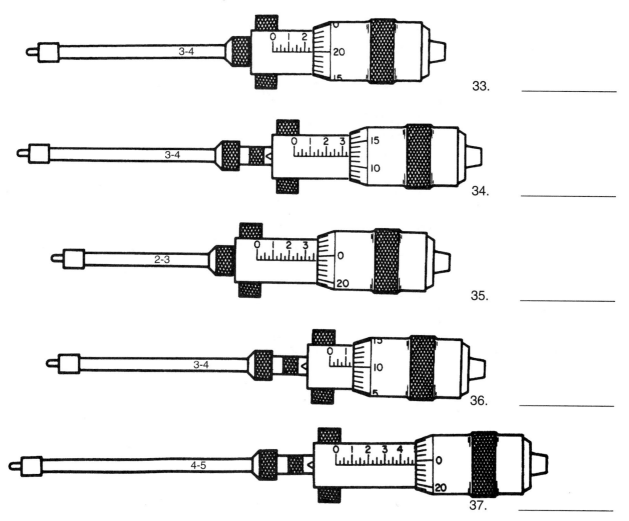

33. _____

34. _____

35. _____

36. _____

37. _____

In problems 38 through 42, demonstrate your ability to read the metric micrometers shown.

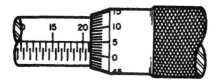

38. _____

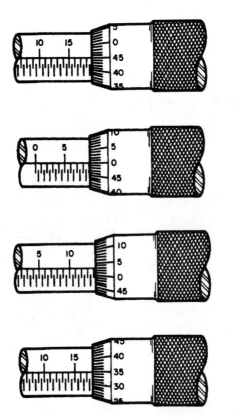

39. _____

40. _____

41. _____

42. _____

Common Fractions

Unit 11 MULTIPLICATION OF COMMON FRACTIONS

COMMON FRACTIONS

There are five terms you should know:

Numerator: The top number of a fraction, the number above the bar.

Denominator: The bottom number of a fraction, the number below the bar. Many students remember the difference between *numerator* and *denominator* by associating the "d" in denominator with "down."

Proper fraction: The type of fractions you are most likely to see: ½, ⅓, ¼, ⅞, ⅘. The numerator (top number) is smaller than the denominator (bottom number).

Improper fraction: Fractions in which the numerator (top number) is bigger than the denominator (bottom number): ⁵⁄₄, ⁸⁄₇, ³⁄₂.

Mixed number: A "mixture" of a whole number and a proper fraction: 2½, 3½, 4⅞.

There are three skills you need to know before working with fractions:

Reducing a Fraction

Unless otherwise stated, all fractions should be reduced to lowest terms. You reduce fractions by dividing both the numerator and the denominator by the same number that is common to both. This is repeated until the fraction can no longer be reduced. For example, to reduce ⁴⁄₁₀, find a single number that is common to both 4 and 10, a number that can go into 4 and 10 evenly. In this case, the number is 2:

Example: $\dfrac{4}{10} \rangle \div 2 = \dfrac{2}{5}$

A single number that goes
evenly into both the numerator
and denominator.

To reduce $^{12}/_{24}$, find a single number that goes evenly into both the numerator and the denominator. There are a few numbers that will work, such as 2, 3, and 12. Pick the greatest number, 12.

Example: $\dfrac{12}{24} = \dfrac{1}{2}$

What happens if we choose 2 or 3? We would reduce to the *same answer,* but we would have to repeat our reducing step until finally arriving at $^{1}/_{2}$.

Changing Improper Fractions into Mixed Numbers

Step 1.   Divide the bottom number into the top number.

Step 2.   Write the remainder as a fraction over the original denominator.

Step 3.   Reduce the fraction, if necessary.

For example, to change $^{8}/_{7}$ into a mixed number, you divide 8 by 7 and write the remainder as a fraction, using the remainder (1) as the numerator and the original denominator (7) as the new denominator:

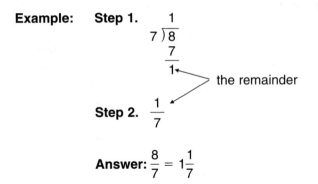

Example: **Step 1.**
$$7\overline{)8}$$

the remainder

Step 2. $\dfrac{1}{7}$

Answer: $\dfrac{8}{7} = 1\dfrac{1}{7}$

Reducing (step 3) is not necessary this time.

Changing Mixed Numbers into Improper Fractions

This process is the opposite of the one above. It is necessary to learn this process because it is the only way to multiply and divide fractions.

Step 1.   Multiply the denominator of the proper fraction by the whole number.

Step 2.   Add the result to the numerator.

Step 3.   Place the number *over* the original denominator.

For example, to change $2^{1}/_{2}$ into a mixed number, multiply the denominator (2) times the whole number (2), add the result (4) to the numerator (1), and place the answer from step 2 (5) over the original denominator (2):

Example: **Step 1.** $2 \times 2 = 4$

   Step 2. $4 + 1 = 5$

   Step 3. $\dfrac{5}{2}$

   Answer: $2\dfrac{1}{2} = \dfrac{5}{2}$

BASIC PRINCIPLES OF MULTIPLICATION OF COMMON FRACTIONS

Step 1.   Multiply the numerators.

Step 2.   Multiply the denominators.

Step 3.   Write the product of the numerators over the product of the denominators.

Step 4.   Reduce the fraction to the lowest terms.

Example: Multiply $\dfrac{3}{4}$ by $\dfrac{1}{8}$ $\dfrac{3 \times 1}{4 \times 8} = \dfrac{3}{32}$

The process can be simplified by *cancellation.* Before multiplying, divide the numerator and the denominator by a number that is common to both.

Example: $\dfrac{4}{16} \times \dfrac{3}{8}$ 4 and 8 or 4 and 16 will cancel to a lower number.

   $\dfrac{\overset{1}{4}}{16} \times \dfrac{3}{8_2} = \dfrac{3}{32}$

To multiply fractions by whole or mixed numbers: $1\frac{1}{2} \times 8$.

Step 1.   Change all mixed numbers to improper fractions by multiplying the denominator by the whole number and adding it to the numerator. For example:

   $2 \times 1 = 2$

   $2 + 1 = 3$

   $1\dfrac{1}{2} = \dfrac{3}{2}$

Step 2.   Change all whole numbers to fractions by placing the whole number over one. For example:

   8 as a fraction would be $\dfrac{8}{1}$

Step 3.   Multiply the numerators.

Step 4.   Multiply the denominators.

Step 5.   If necessary, reduce the answer to lowest terms.

Example: $\dfrac{9}{16} \times 1\dfrac{1}{2}$

$$1\dfrac{1}{2} = \dfrac{3}{2} \text{ (Step 1)}$$

$$\dfrac{9}{16} \times \dfrac{3}{2} = \dfrac{27}{32} \text{ (Steps 3 and 4)}$$

Answer: $\dfrac{9}{16} \times 1\dfrac{1}{2} = \dfrac{27}{32}$

PRACTICAL PROBLEMS

Reminder: Reduce all fractions to lowest terms in the final answer.

1. How many pounds of grease are there in a barrel that holds about 54 gallons? (Use $7\frac{7}{8}$ pounds per gallon.)   _____

2. What is the weight of the grease in a 216-liter drum if there are about $3\frac{1}{2}$ kilograms per liter?   _____

3. A mechanic cuts 12 pieces of copper tubing from a coil. Each piece is $37\frac{7}{16}$ inches long. What is the total length used?   _____

4. A mechanic cuts 5 bushings from a piece of stock. Each bushing is $\frac{7}{8}$ inch long. Not allowing for saw cuts, what is the total length used?   _____

5. If one piston weighs $\frac{3}{4}$ pound, what do 8 pistons weigh?   _____

6. A gasoline tank holds 60 liters. How many kilometers is it possible to travel on one tank of gasoline at $8\frac{6}{10}$ kilometers per liter?   _____

7. A car averages $18\frac{1}{2}$ miles to a gallon of gasoline. How many miles can be traveled on 37 gallons?   _____

8. A man drives a car an average of $275\frac{1}{2}$ miles a day. How many miles are driven in $4\frac{3}{4}$ days?   _____

9. An auto travels a mile in 49 seconds. How long does it take to travel $\frac{5}{8}$ mile at the same rate?   _____

10. An auto averages $76\frac{1}{2}$ kilometers per hour on a trip. How far does it go in $7\frac{3}{4}$ hours?   _____

Note: Use this diagram for problems 11 through 13.

On a 16 NF thread the nut advances $\frac{1}{16}$ inch per revolution;
on a 28 NF it advances $\frac{1}{28}$, inch, etc.

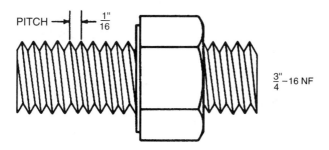

11. How far will a nut advance with 14 turns on a $\frac{1}{4}$-inch–28 NF (National Fine
     Thread) bolt?   _____

12. How far will a nut advance with 15 turns of a $\frac{1}{4}$-inch–20 NC (National Coarse
     Thread) bolt?   _____

13. How far will a nut advance with 6 turns on a $\frac{3}{4}$-inch–16 NF bolt?   _____

14. A certain car (four-door) requires rubber door weather strip material as
     follows:

     Each rear door–$134\frac{7}{8}$ inches
     Each front door–$138\frac{3}{4}$ inches

     a. How many inches of weather strip material are needed for two rear
        doors?   a. _____

     b. How many inches of weather strip material are needed for two front
        doors?   b. _____

     c. How many inches of weather strip material are needed for all
        four doors?   c. _____

15. A certain car requires $22\frac{3}{16}$ inches of $\frac{13}{32}$-inch air-conditioning hose. How
     many inches of hose are needed for four cars?   _____

16. Each power steering unit needs $9\frac{3}{4}$ inches of power steering return hose.
     How many inches of hose are needed to replace the hose in eight units?   _____

17. The specifications of rubber door weather strip for a certain car are: 4 pieces,
     each $123\frac{1}{4}$ inches long; and 4 pieces, each $130\frac{1}{2}$ inches long. What is the
     total length needed in inches?   _____

18.   A certain car requires 63¾ centimeters of 1¼-centimeter air-conditioning hose. How many centimeters are needed to replace the hose in four cars?   _____

19.   Five lengths of gas-line hose, each 2⅞ inches long, are cut from a piece of hose 24½ inches long. How many inches of hose are left?   _____

20.   A channel iron crossmember weighs 2½ pounds per foot. What is the weight of five crossmembers, each 2½ feet long?   _____

21.   Tubular steel crossmembers weigh 3⅛ pounds per foot. What is the weight of two of these members, one 2 feet long and one 2½ feet long?   _____

22.   The cooling system of a Chevrolet 6-cylinder 250-cubic-inch-displacement (CID) engine has a capacity of 3½ gallons. A solution that is ³⁄₇ ethylene glycol offers winter protection against freezing. How many quarts of ethylene glycol are used?   _____

23.   The capacity of a cooling system is 3¾ gallons. The owner wants protection against freezing to a temperature of −43°F. If 17 parts antifreeze and 15 parts water give this protection, how many quarts of antifreeze are used?   _____

   Note: Use this diagram for problems 24 through 26.

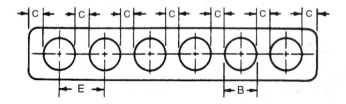

24.   On the cylinder block surface the cylinders are 3¹⁄₁₆ inches in diameter, and distance **C** is ⅝, inch. Find, in inches, the length of the block.   _____

25.   On the cylinder block, dimension **B** is 3¹⁄₁₆ inches, and dimension **C** is ⅝ inch. What is the center-to-center length of dimension **E**?   _____

26.   The cylinders are 3³¹⁄₃₂ inches in diameter and distance **C** is ⅞ inch. What is the total length of the cylinder block surface?   _____

Note: Use this diagram for problems 27 and 28.

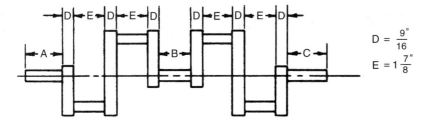

$D = \dfrac{9"}{16}$

$E = 1\dfrac{7"}{8}$

27. What is the length of this crankshaft if main bearing journals **A, B,** and **C** are each $1\frac{5}{8}$ inches long?   _____

28. The illustrated crankshaft is $15\frac{3}{8}$, inches long, and the main bearing journals **A** and **C** are $1\frac{7}{16}$ inches long. What is the length of the center main bearing journal **B**?   _____

29. Twelve $\frac{7}{16}$-inch diameter holes are to be drilled in a straight line. A distance of $\frac{3}{4}$ inch is left between the edges of the holes and at each end. What length piece, in inches, is needed?   _____

30. Nine holes are drilled in a straight line with a center-to-center distance of $2\frac{9}{10}$ inches. The distance allowed between each end and the center of the end holes is $6\frac{6}{10}$ inches. What length piece, in inches, is needed?   _____

31. A certain car averages $33\frac{2}{5}$ miles per gallon of gas. How far will it travel on $21\frac{1}{2}$ gallons of gas?   _____

Unit 12 DIVISION OF COMMON FRACTIONS

BASIC PRINCIPLES OF DIVISION OF COMMON FRACTIONS

Step 1.   Invert the fraction to the *right* of the division sign. *Invert* means to place the bottom number on top and the top number on the bottom.

Step 2.   Change the division sign to a multiplication sign and proceed as in multiplication of common fractions.

Example: $\dfrac{7}{8} \div \dfrac{1}{4}$ (divisor)

$$\dfrac{7}{8} \times \dfrac{4}{1} \longleftarrow \text{Steps 1 and 2}$$

$$\longleftarrow \text{Divide 7 by 2}$$

$$\dfrac{7}{8_2} \times \dfrac{4^1}{1} = \dfrac{7}{2} = 3\dfrac{1}{2} \longleftarrow \text{Answer is a mixed number}$$

Dividing fractions and mixed and/or whole numbers:

Step 1.   Change all mixed numbers to improper fractions.

Step 2.   Change all whole numbers to fractions.

Step 3.   Invert the divisor.

Step 4.   Cancel common factors in both numerator and denominator.

Step 5.   Multiply the fractions.

Step 6.   Reduce to lowest terms, if necessary.

Example: $6\dfrac{9}{16} \div 5$

Step 1. $6\dfrac{9}{16} = \dfrac{105}{16}$

Step 2. $5 = \dfrac{5}{1}$

Step 3. $\dfrac{105}{16} \times \dfrac{1}{5}$

Step 4. $\dfrac{\overset{21}{105}}{16} \times \dfrac{1}{\underset{1}{5}}$

Step 5. $\dfrac{21 \times 1}{16 \times 1} = \dfrac{21}{16} = 1\dfrac{5}{16}$

Answer: $6\dfrac{9}{16} \div 5 = 1\dfrac{15}{16}$

PRACTICAL PROBLEMS

Note: Use this diagram for problems 1 through 3.

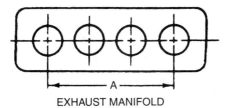

EXHAUST MANIFOLD

1.   The holes in the sketch are equally spaced, and **A** is 21¾ inches. What is the distance (center-to-center) between the holes?   _____

2.   If **A** is 18³⁄₃₂ inches, what is the center-to-center distance between holes?   _____

3.   If **A** is 12¾ inches, what is the center-to-center distance between holes?   _____

Note: Use this diagram for problems 4 and 5.

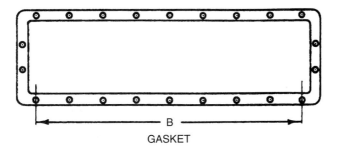

GASKET

4.   The gasket holes are equally spaced, and **B** is 22½ inches. What is the center-to-center distance between holes?   _____

5.   If, in the gasket, distance **B** is 16⅞ inches, what is the center-to-center distance between holes?   _____

6.   The average time for lubricating a car is ⅔ hour. How many cars can be lubricated in an 8-hour day?   _____

7.   The spark plugs are replaced in 12 cars in 10½ hours. What is the average time spent on each car?   _____

8.   A certain size copper tubing weighs ⅓ pound per foot. How many feet are there in a roll weighing 20¼ pounds?   _____

9.   A garage owner buys a 50-foot roll of ⅝-inch heater hose. How many cars can be repaired if 9½ feet of hose are required for each car?   _____

10. A car uses 175½ inches of TVRS wire for 6 spark plugs. What is the average wire length to each spark plug? _____

11. An automobile jack raises the car ³⁄16 inch for each stroke of the lever. How many strokes of the lever are needed to lift the car 3¾ inches? _____

12. A garage owner has an oil barrel that contains 53¾ gallons. If sales average 10¾ gallons a day, how many days will the oil last? _____

13. A customer drives a car 120 miles and uses 3¾ gallons of gas. How many miles does the car average per gallon of gas? _____

14. At an odometer reading of 100 miles, the gasoline gauge on a car registers 3½ gallons. At the odometer reading of 1,339 miles, the gauge reads 4½ gallons. The owner has had 45 gallons of gasoline placed in the tank during this period. What is the average number of miles per gallon this car has traveled? _____

15. At an average rate of 33½ miles per hour, how many hours does it take to drive 410⅜ miles? _____

16. How many 4⅞-inch lengths of air duct tubing can be cut from a 3-foot piece of 1⁷⁄16-inch hose? _____

17. A mechanic is paid $55.80 for 4½ hours of work. What is the hourly rate in dollars and cents? Note: $55.80 equals 5,580 cents. _____

18. How many times is a nut revolved to take up ½" on a ¼"–28 NF bolt? _____

19. How many times is a nut revolved to take up ½" on a ¼"–20 NC bolt? _____

20. Find the number of turns necessary to take up ⅜" on a ¾"–16 NF bolt. _____

Note: Use this diagram for problems 21 and 22.

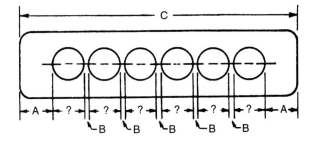

21. Distance **A** is 2¼ inches, distance **B** is 1⅜ inches, and distance **C** is 36⅛ inches. What is the diameter of each cylinder bore? _____

22. If **A** is 2½ inches, **B** is 1⅛ inches, and **C** is 30⅛ inches, what diameter is the cylinder bore? _____

23. When taking stock inventory, a mechanic finds that 25 of the ¼"–20 nuts weigh 15 ounces. The total weight of all the ¼"–20 nuts is 5 pounds 13 ounces. How many nuts are there? _____

Note: Use this diagram for problems 24 and 25. Stroke is equal to 2 × the throw.

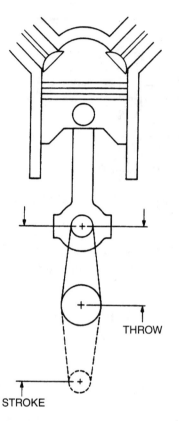

THROW

STROKE

24.   What is the crank throw when the stroke is 4³⁄16 inches?   _____

25.   What is the crank throw when the stroke is 3¾ inches?   _____

26.   How many ⅞-inch pieces can be cut from a piece of safety wire 7 inches long?   _____

27.   How many pieces of 1⁵⁄16 inches long *#16* wire can be cut from a piece 13⅛ inches long?   _____

28.   How many 7⅜-inch pieces of copper tubing can be cut from a 10-foot length?   _____

29.   A lot of bushings costs $22.84. If the price per bushing is 65¼ cents, how many bushings are in the lot? Note: $22.84 equals 2,284 cents.   _____

30.   Two 10-foot lengths of wire are cut up into pieces 6¾ inches long. How many 6¾-inch pieces are made?   _____

31. How many $3\frac{7}{8}$-inch long metal shims can be made from 20 metal strips, each 3 feet long? _____

32. If $9\frac{2}{3}$ yards of gasket material cost $58.00, how much does it cost per yard? Note: $58.00 equals 5,800 cents. _____

33. How many $16\frac{3}{8}$-inch lengths can be cut from a 10-foot length of wire? _____

34. If $7\frac{3}{4}$ dozen wire terminals cost $32.55, what is the cost per dozen? Note: $32.55 equals 3,255 cents. _____

35. How many $1\frac{3}{4}$-inch machine bolt blanks can be cut from a 5-foot length of stock? Allow $\frac{7}{32}$ inch for waste on each blank. _____

36. If a woman averages $45\frac{3}{4}$ miles per hour on a trip, how long does it take her to travel $571\frac{7}{8}$ miles? _____

37. A lot of bolts costs $25.25. At $25\frac{1}{4}$ cents each, how many bolts are there in the lot? Note: $25.25 equals 2,525 cents. _____

Unit 13 ADDITION OF COMMON FRACTIONS

BASIC PRINCIPLES OF ADDITION OF COMMON FRACTIONS

Accurate measurement in the automotive trade requires closer measurement than can be done with whole numbers. This can be done by dividing the inch into many equal parts, called common fractions: ½, ¼, ⅛, ¹⁄₁₆, ¹⁄₃₂, and ¹⁄₆₄.

The two parts of a fraction are the numerator (above the line) and denominator (below the line).

To add or subtract fractions, you must first find the least common denominator (LCD). The least common denominator is the *smallest number that all of the denominators will divide into.*

Example: $\dfrac{3}{32}, \dfrac{1}{8}, \dfrac{1}{16}, \dfrac{5}{64},$ $32\overline{)64}\;^{2}$ $8\overline{)64}\;^{8}$ $16\overline{)64}\;^{4}$ $64\overline{)64}\;^{1}$

LCD is 64

Step 1. Divide the LCD by the denominator of each fraction.

Step 2. Multiply both the numerator and denominator by this number.

Example:

$$\dfrac{3 \times 2 = 6}{32 \times 2 = 64}$$

$$\dfrac{1 \times 8 = 8}{8 \times 8 = 64}$$

$$\dfrac{1 \times 4 = 4}{16 \times 4 = 64}$$

$$+\dfrac{5 \times 1 = 5}{64 \times 1 = 64}$$

Answer: $\dfrac{23}{64}$

It is sometimes necessary to reduce the answer to its lowest terms. This is done by dividing both numerator and denominator by the same number. This is repeated until the fraction can no longer be reduced.

ADDING COMMON FRACTIONS, MIXED NUMBERS, AND WHOLE NUMBERS

A mixed number is a whole number and a fraction.

Example: $3\frac{1}{4}$, $5\frac{1}{8}$, $4\frac{5}{16}$

To add a group of fractions, mixed numbers, and whole numbers, follow these steps:

Step 1. Place all the units to be added in a vertical column.

Step 2. Find the LCD for the fractions.

Step 3. Add the numerators (above the line).

Step 4. Put the sum above the LCD.

Step 5. If this sum is larger than the LCD, divide it by the LCD.

Step 6. Add the whole number.

Example: $5\frac{1}{2} = 3\frac{1}{4} + \frac{9}{16} + \frac{1}{8}$

Step 1.
$$5\frac{1}{2}$$
$$3\frac{1}{4}$$
$$\frac{9}{16}$$
$$\frac{1}{8}$$

Step 2.
$$5\frac{1}{2} \times \frac{8}{8} = 5\frac{8}{16}$$
$$3\frac{1}{4} \times \frac{4}{4} = 3\frac{4}{16}$$
$$\frac{9}{16} \times \frac{1}{1} = \frac{9}{16}$$
$$\frac{1}{8} \times \frac{2}{2} = \frac{2}{16}$$

Step 3.

Step 4. $\frac{23}{16} =$ **Step 5.** $1\frac{7}{16}$

Step 6.

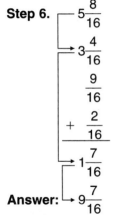

$$5\frac{8}{16}$$
$$3\frac{4}{16}$$
$$\frac{9}{16}$$
$$+ \frac{2}{16}$$
$$1\frac{7}{16}$$

Answer: $9\frac{7}{16}$

PRACTICAL PROBLEMS

Note: Use this diagram for problems 1 through 4.

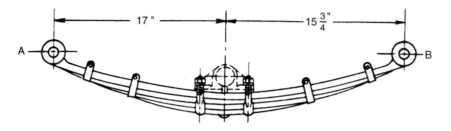

1.   What is the length of the spring, in inches, from the center of eye **A** to the center of eye **B**?   _____

2.   If the 17-inch dimension is changed to $16\frac{3}{4}$ inches, what is the length of the spring in inches?   _____

3.   If the measurements are given as $17\frac{5}{8}$ inches and $14\frac{7}{8}$ inches, what is the spring length in inches?   _____

4.   If the measurements are given as $42\frac{1}{2}$ centimeters and $38\frac{1}{5}$ centimeters, what is the spring length in centimeters?   _____

5.   A mechanic works $2\frac{1}{4}$ hours on one car, $1\frac{1}{2}$ hours on another car, and $5\frac{3}{4}$ hours on a third car. What is the total number of hours worked?   _____

6.   A garage owner has 4 pieces of vacuum hose in lengths of $9\frac{3}{4}$ feet, $10\frac{1}{4}$ feet, $49\frac{1}{2}$ feet, and 12 feet. How many feet of vacuum hose are in stock?   _____

7.   In one day a truck hauls loads of $3\frac{1}{4}$ tons, $4\frac{1}{8}$ tons, and $3\frac{1}{2}$ tons of stone. What is the total tonnage carried during that day?   _____

8.   The chassis of a truck is $14\frac{3}{4}$ feet long. At the rear, the body projects $4\frac{1}{2}$ feet beyond the end of the frame. There is clearance space of $3\frac{3}{4}$ feet at each end of the truck when it is in the garage. What is the inside length of the garage in feet?   _____

9. In the drawing, what is the thickness, in inches, of the crossmember and frame at the point where the hole is drilled? _____

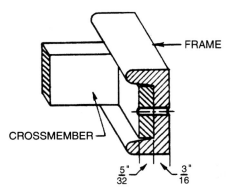

FRAME

CROSSMEMBER

$\frac{5"}{32}$ $\frac{3"}{16}$

10. If a frame is $^7\!/_{16}$ inch thick and the crossmember is $^5\!/_{32}$ inch thick, what is the combined thickness of both members? _____

11. In a car the crossmember is $^1\!/_8$ inch thick, and the frame is $^3\!/_{32}$ inch thick. How much stock is drilled to pierce both pieces? _____

12. A truck frame is $^3\!/_{16}$ inch thick. The crossmember is $^{13}\!/_{64}$ inch thick. How much stock is drilled to pierce both pieces? _____

Note: Use this diagram for problems 13 through 15.

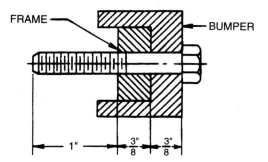

FRAME   BUMPER

1" $\frac{3"}{8}$ $\frac{3"}{8}$

13. In the illustration, what is the under-the-head length of bolt used in inches? _____

14. If the same bumper is used but there is only $^{13}\!/_{16}$ inch projecting beyond the frame, what length bolt, in inches, is needed? _____

15. A bumper is $^7\!/_{16}$ inch thick, the frame is $^1\!/_4$ inch thick, and there is $^5\!/_8$ inch projecting. What length bolt (under-the-head), in inches, is needed? _____

Note: Use this diagram for problems 16 through 18.

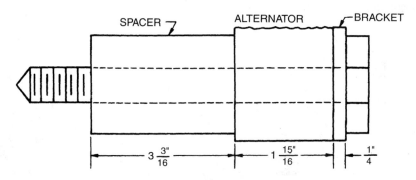

16.   To mount the alternator on a Chevrolet V-8, the bolt goes through a bracket, the alternator, and a spacer before it goes into the cylinder head. What is the distance, in inches, from the underside of the bolt head to the cylinder head?   _____

17.   After the bolt goes through the bracket, alternator, and spacer, ⅝" of thread is required to hold it securely to the cylinder head. Using the diagram, find the length of the bolt needed.   _____

18.   The lengths of the spacer and the alternator in the diagram are changed to 3¹⁄₁₆" and 2¹⁄₁₆", respectively. What is the distance in inches, from the underside of the bolt head to the cylinder head?   _____

Note: Use this diagram for problems 19 and 20.

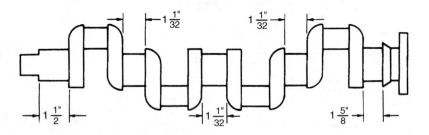

19.   What is the total length, in inches, of the five main bearing journals in this crankshaft?   _____

20.   The three center journals are 1³⁄₃₂ inches long, the front journal is 1⁷⁄₁₆ inches long, and the rear journal is 1⁹⁄₁₆ inches long. What is the total length of the five main bearing journals?   _____

21. An interior refinishing job calls for these lengths of simulated wood grain decal material: 2 pieces, $21\frac{1}{2}$ centimeters each; 2 pieces, $18\frac{3}{4}$ centimeters each; 2 pieces, $93\frac{3}{4}$ centimeters each; and 2 pieces, $70\frac{1}{4}$ centimeters each. What is the total length, in centimeters, needed for this job?   _____

22. The interior of a van is $5\frac{1}{4}$ feet wide by $11\frac{3}{4}$ feet long. The owner buys carpeting in a 3-foot wide strip. The amounts needed are: 2 strips, $11\frac{3}{4}$ feet long for the floor; 2 strips, $10\frac{1}{2}$ feet long for the sides; and 1 strip, $5\frac{1}{4}$ feet long for part of the doors in back. How many feet of carpeting should the owner buy?   _____

23. A repairer uses lengths of copper tubing for oil lines as follows: $12\frac{7}{32}$ inches, $14\frac{9}{64}$ inches, $9\frac{7}{16}$ inches, 4 inches, $7\frac{1}{8}$ inches, and $6\frac{7}{8}$ inches. What is the total length, in inches, of tubing used?   _____

24. A piece of hose has an inside diameter of $1\frac{7}{8}$ inches and a wall thickness of $\frac{5}{32}$ inch. What is the outside diameter in inches?   _____

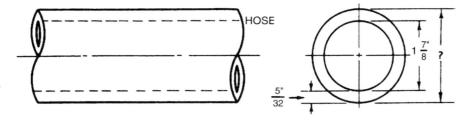

25. Find the outside diameter of a piece of hose with a $1\frac{3}{4}$-inch inside diameter and a wall thickness of $\frac{7}{32}$ inch.   _____

26. What is the outside diameter of a piece of copper tubing with an inside diameter of $\frac{5}{16}$ inch and a wall thickness of $\frac{3}{32}$ inch?   _____

27. A pipe has a $\frac{1}{2}$-inch wall thickness and a $3\frac{3}{4}$-inch inside diameter. Find the outside diameter of the pipe.   _____

28. A washer has a ¾-inch hole and ¹³⁄₃₂ inch between the hole and outside edge. What is the outside diameter of the washer?   _____

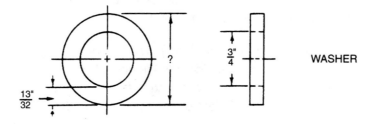

29. A washer has a ¹⁵⁄₁₆-inch hole and ¹¹⁄₃₂ inch between the hole and outside on each side. What is the outside diameter?   _____

Note: Use this diagram for problems 30 and 31.

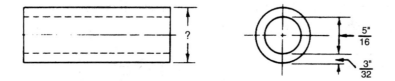

30. A shaft, ⁵⁄₁₆ inch in diameter, is to run in a bronze bushing with a wall thickness of ³⁄₃₂ inch. What size hole is drilled for the bushing?   _____

31. If a shaft is ½ inch in diameter and a bronze bushing is ⁵⁄₃₂ inch thick, what size hole is drilled for the bushing?   _____

Note: Use this diagram for problems 32 through 35.

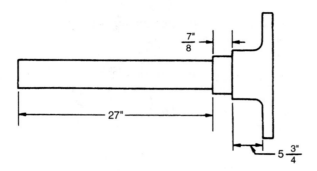

32. What is the overall length of the axle shaft?   _____

33. If the 27-inch dimension of the axle shaft shown is changed to 16⁹⁄₁₆ inches, what is the overall length?   _____

34. An axle shaft has dimensions of 21 inches, $^{15}\!/_{16}$ inch, and $7^7\!/_8$ inches. How long is the axle shaft? _____

35. How long is the axle shaft with dimensions of $30^{13}\!/_{16}$ inches, $1^9\!/_{32}$ inches, and $6^1\!/_2$ inches? _____

36. What length piece is used to make this drill and tap block? _____

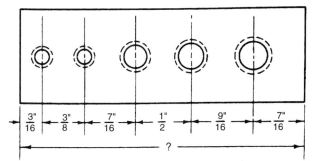

37. A mechanic needs the following lengths of $^3\!/_8$-inch copper tubing: $14^1\!/_2$ inches, $7^3\!/_8$ inches, $11^{15}\!/_{16}$ inches, and $16^{15}\!/_{32}$ inches. How much tubing is needed in all? _____

38. A mechanic works $2^1\!/_4$ hours on one car, $5^1\!/_2$ hours on another, and 3 hours on a third. How many hours are spent on the three cars? _____

Note: Use this diagram for problems 39 and 40.

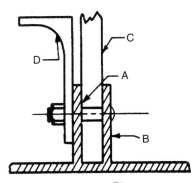

39. In the illustrated assembly, **A** and **B** are $^7\!/_{32}$ inch thick, **C** is $1^1\!/_2$ inches thick, and **D** is $^5\!/_{32}$ inch thick. An allowance is made for a nut $^7\!/_{16}$ inch thick. What length bolts are purchased to hold the assembly together? _____

40. How long are the bolts in the illustrated assembly if **A** and **B** are $^3\!/_8$ inch thick, **C** is $1^3\!/_{16}$ inches thick, **D** is $^3\!/_{16}$ inch thick, and the nut is $^3\!/_8$ inch thick? _____

41.   In rewiring the ignition system of a compact foreign car, an electrical systems
      specialist uses the following lengths of high-tension wire: $45\frac{1}{2}$ centimeters,
      $54\frac{7}{10}$ centimeters, 45 centimeters, $48\frac{4}{10}$ centimeters, and $60\frac{1}{2}$ centimeters.
      What is the total length of wire used in centimeters?   _____

Unit 14 SUBTRACTION OF COMMON FRACTIONS

BASIC PRINCIPLES OF SUBTRACTION OF COMMON FRACTIONS

Review Unit 13 for an explanation of the least common denominator (LCD).

Step 1. Find the LCD.

Step 2. Change all fractions using the LCD.

Step 3. Subtract the numerators.

Step 4. Place the remainder *over* the LCD to give the answer as a fraction.

Example:

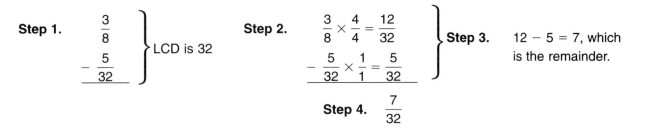

Subtracting a fraction from a whole number:

Example: Subtract $\frac{3}{8}$ from 7.

Step 1. Arrange vertically:

$$
\begin{array}{r}
7 \\
-\ \frac{3}{8} \\
\hline
\end{array}
$$

The goal is to be able to subtract $\frac{3}{8}$ from 7, but there are no fractions with the whole number, so we must rewrite the seven as a whole number *and* a fraction, in other words, a mixed number.

Step 2. Change the whole number into a fraction by *borrowing* one from it, making it 6.

Step 3. Take the one that was borrowed and rewrite it as a fraction with the *same top* and *bottom* number as the denominator in the other fraction. In other words, borrow one from the whole number and rewrite it as a fraction *equal to one,* which means that the fraction will have the same numerator and denominator. The numbers that we use to form the new fraction depend on the *denominator* of the other fraction in the problem; therefore, 7 becomes $6\frac{8}{8}$ which is equal to 7.

$$6\frac{8}{8}$$
$$-\ \frac{3}{8}$$

Step 4.   Subtract the numerators, $8 - 3 = 5$, and bring down the denominator. The whole number 6 is also brought down in this case because it is regular subtraction: $6 - 0 = 6$. Reduce the fraction if necessary.

Answer: $6\frac{5}{8}$

Subtracting a mixed number from a whole number:

Example: Subtract $1\frac{5}{8}$ from 9.

Step 1.   Subtract one from the number $9 - 1 = 8$.

Step 2.   Change the one to a fraction having the same denominator as the fraction in the mixed number: $1 = \frac{8}{8}$

$$8\frac{8}{8} - 1\frac{5}{8}$$

Step 3.   Subtract the fractions.

Step 4.   Subtract the whole numbers.

Step 5.   If necessary, reduce to lowest terms.

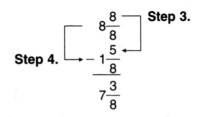

Subtracting a mixed number from a mixed number:

Example: Subtract $4\frac{3}{8}$ from $6\frac{3}{16}$.

Step 1.   Find the LCD for the fractions: $4\frac{3}{8} = 4\frac{6}{16}$.

Step 2.   If the fraction in the larger number is *not* larger than the other fraction, borrow one from the whole number and change it to a fraction.

Step 3.   Add the two fractions: the original $\frac{3}{16}$ plus the $\frac{16}{16} = \frac{19}{16}$.

Step 4.   Subtract the fractions: $\frac{19}{16} - \frac{6}{16} = \frac{13}{16}$.

Step 5.   Subtract the whole numbers: $5 - 4 = 1$.

Step 6.   If necessary, reduce to lowest terms.

Step 1.
$$6\frac{3}{16} \times \frac{1}{1} = 6\frac{3}{16}$$
$$-4\frac{3}{8} \times \frac{2}{2} = 4\frac{6}{16}$$

Steps 2 and 3.
$$5\frac{16}{16} + \frac{3}{16} = 5\frac{19}{16}$$

Step 5. $\qquad$ **Step 4.**
$$5\frac{19}{16}$$
$$-4\frac{6}{16}$$

Answer: $1\frac{13}{16}$

PRACTICAL PROBLEMS

1. The outside diameter of a hose is $2\frac{1}{4}$ inches, and the wall is $\frac{13}{32}$ inch thick. How large, in inches, is the inside diameter?   _____

2. A piece of $\frac{1}{4}$-inch copper tubing measures $\frac{9}{16}$ inch outside. What is the difference between the inside and outside diameters in inches?   _____

3. The outside diameter of a hose is $1\frac{11}{16}$ inches, and the wall is $\frac{3}{16}$ inch thick. What is the size of the inside diameter in inches?   _____

Note: Use this diagram for problems 4 through 8.

LOCKWASHER

4. The lockwasher used on a ⅜-inch bolt has a $^{25}\!/_{64}$-inch inside diameter and a ⅝-inch outside diameter. What is the difference between the inside and outside diameters?   _____

5. What is the clearance between the lockwasher with a $^{25}\!/_{64}$-inch inside diameter and a ⅜-inch diameter bolt?   _____

6. The lockwasher used on a ½-inch bolt has a $^{27}\!/_{32}$-inch outside diameter, and the ring is $^{3}\!/_{16}$ inch in width. What is the measurement, in inches, of the inside diameter?   _____

7. The lockwasher used on a ¾-inch bolt has a $1^{13}\!/_{16}$-inch outside diameter, and the ring is $^{5}\!/_{32}$ inch in width. What is the inside diameter?   _____

8. A lockwasher has a $1^{3}\!/_{64}$-inch outside diameter and a ring width of $^{3}\!/_{16}$ inch. What is the clearance between the lockwasher and a ⅝-inch bolt?   _____

9. A shaft with a $^{3}\!/_{32}$-inch bushing measures $1^{3}\!/_{8}$ inches. What is the diameter of the shaft, in inches, without the bushing?   _____

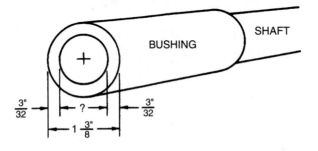

Note: Use this diagram for problems 10 and 11.

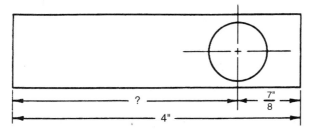

10.   What is the missing dimension in this drawing?   _____

11.   If the length of stock is 5½ inches instead of 4 inches, what is the missing dimension in this drawing?   _____

12.   A piece of stock 6¾ inches long has a hole drilled with its center 2⁵⁄₁₆ inches from one end. How far is it from the center of the hole to the other end of the stock?   _____

13.   What is the length of the short end on an automobile leaf spring if the complete spring is 32 inches long and the long end is 18⅜ inches?   _____

14.   A cooling system contains 14½ liters of water. How many liters of water remain in the system after 3½ liters are drawn out?   _____

15.   An automobile driver plans to drive 325¼ miles in a day. By the middle of the afternoon, 267¾ miles have been traveled. How many miles remain to be driven?   _____

16.   A drum of grease complete with container weighs 420½ pounds. If the container weighs 46¼ pounds, how much does the grease weigh?   _____

17.   A mechanic needs an axle shaft key ¼ inch square. The only available piece of key stock measures ¼ inch × ⅜ inch. How many inches of stock have to be removed to bring it to proper size?   _____

18.   The front wheels of a car are not parallel but set to toe-in. *Toe-in* means that the front edges are closer together than the back. The toe-in measurement of a certain car is ⅜ inch. The specifications call for a toe-in of ³⁄₁₆ inch. By how much should the toe-in be increased or decreased?   _____

19.   A mechanic needs 6 bolts ⁵⁄₁₆ inch in diameter and ⅝, inch long. The only size available is ⁵⁄₁₆ inch in diameter by 1¼ inches long. How much is cut off of each bolt?   _____

Note: Use this diagram for problems 20 and 21.

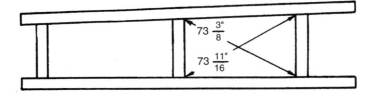

$73\frac{3}{8}"$

$73\frac{11}{16}"$

20.   While checking a frame for alignment a mechanic finds that one diagonal measures $73\frac{11}{16}$ inches and the other measures $73\frac{3}{8}$ inches. What is the difference in the measurements taken?

_____

21.   If the measurements of the diagonals for alignment are $87\frac{1}{16}$ inches and $86\frac{9}{32}$ inches, what is the difference?

_____

Note: Use this diagram for problems 22 and 23.

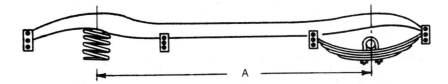

A

22.   The front axle of a pickup is to be aligned with the rear axle. On one side the measurement **A** is 9 feet $3\frac{3}{4}$ inches, and on the other side the measurement is 9 feet $4\frac{5}{16}$ inches. How much does one side need to be shifted to bring the axles into alignment?

_____

23.   Measurement **A** is $108\frac{7}{32}$ inches on one side of the car and $109\frac{3}{16}$ inches on the other side. How much out of alignment are the two axles?

_____

24.   The long end of a spring is $53\frac{1}{2}$ centimeters long. How long, in centimeters, is the short end of the spring if the overall length is $83\frac{3}{4}$ centimeters?

_____

25.   Two splice-plates are cut from a piece of sheet steel that has an overall length of $15\frac{3}{4}$ inches. The plates are $6\frac{7}{8}$ inches and $7\frac{5}{16}$ inches long, and $\frac{1}{16}$ inch is allowed for each saw cut. How much material is left after the cutting?

_____

26.   An engine specialist works $1\frac{1}{4}$ hours on one car, $3\frac{3}{4}$ hours on another car, and $1\frac{1}{4}$ hours on a third car. How much time remains for another job in an 8-hour working day?

_____

Note: Use this diagram for problems 27 through 31

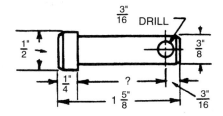

27.   What is the missing dimension, in inches, of this clevis pin?   _____

28.   If the ¼-inch dimension is changed to ⅜ inch, what is the missing dimension of the clevis pin?   _____

29.   If total length is changed to 1¾ inches and the ¼-inch dimension is used, what is the missing dimension of the clevis pin?   _____

30.   What is the difference in size between the two ends of the clevis pin?   _____

31.   How far is the center of the hole from the top (including the head) of the clevis pin?   _____

Note: Use this diagram for problems 32 through 34.

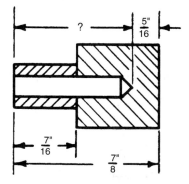

32.   In the sketch, what is the depth of the drilled hole in inches?   _____

33.   If the hole has been drilled ⁷⁄₁₆ inch deep, how much deeper must it be drilled?   _____

34.   If it is necessary to keep the hole ½ inch from the bottom, to what depth can the hole be drilled?   _____

35.   A truck has an overall length of 14$\frac{3}{10}$ meters. A garage is 22$\frac{7}{10}$ meters long inside. How much clearance space is left if a bench 2 meters wide is placed in front of the truck?   _____

36.   The top of the cab of a truck is 8$\frac{1}{2}$ feet from the ground. The top of the body is 3$\frac{3}{4}$ feet above the cab. What clearance is there when going under a bridge that has a clearance of 13$\frac{3}{4}$ feet?   _____

37.   Find, in inches, dimension **F** in the truck wheel hub drawing. **A** = $\frac{1}{16}$ inch, **B** = $\frac{9}{16}$ inch, **C** = $\frac{31}{32}$ inch, **D** = $\frac{5}{16}$ inch, **E** = 4$\frac{5}{32}$ inches.   _____

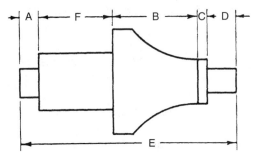

38.   A garage owner has a piece of radiator hose that is 69 inches long. From it, pieces of 5$\frac{1}{8}$ inches, 6$\frac{7}{16}$ inches, 9$\frac{16}{32}$ inches, 4 inches, 5$\frac{7}{8}$ inches, and 15$\frac{3}{8}$ inches are cut. How many inches remain?   _____

39.   The following amounts are taken from a 5-gallon can of oil: $\frac{3}{4}$ gallon, 1$\frac{3}{4}$ gallons, 1$\frac{1}{2}$ gallons, and $\frac{1}{8}$ gallon. How much oil remains?   _____

40.   A gasoline tank contains 18$\frac{1}{2}$ gallons of gasoline. The owner uses 7$\frac{3}{4}$ gallons on one trip and 2$\frac{1}{3}$ gallons on another. How many gallons of gasoline remain in the tank?   _____

41.   Find, in inches, the measurement of **A** in the sketch.   _____

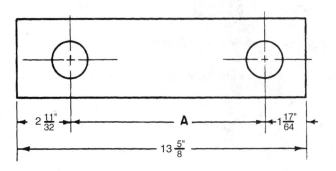

42. A worker is allowed 14$\frac{1}{4}$ hours to complete a job. The five different times already spent are $\frac{1}{2}$ hour, 1$\frac{1}{4}$ hours, 7$\frac{3}{4}$ hours, 1$\frac{1}{4}$ hours, and $\frac{3}{4}$ hour. How much time remains to complete the job? _____

43. A garage owner uses a 7$\frac{1}{2}$-horsepower electric motor to run the air pump and machine shop. If the machine shop requires 4$\frac{1}{2}$ horsepower and the air pump requires $\frac{3}{4}$ horsepower, what amount of power is still available for other uses? _____

The following problems test ability to add and subtract dimensions.

Note: The symbol (") is used to represent inch or inches.

Addition		Subtraction	
44. $\frac{3}{8}$" + $\frac{5}{8}$"	_____	64. 1" − $\frac{3}{8}$"	_____
45. $\frac{1}{4}$" + $\frac{1}{8}$"	_____	65. $\frac{3}{4}$" − $\frac{1}{8}$"	_____
46. $\frac{1}{2}$" + $\frac{3}{8}$"	_____	66. $\frac{1}{2}$" − $\frac{3}{16}$"	_____
47. $\frac{5}{16}$" + $\frac{7}{16}$"	_____	67. 1$\frac{1}{4}$" − $\frac{1}{16}$"	_____
48. $\frac{3}{16}$" + $\frac{1}{4}$"	_____	68. 1$\frac{1}{2}$" − $\frac{5}{16}$"	_____
49. $\frac{7}{8}$" + $\frac{1}{16}$"	_____	69. 2$\frac{3}{4}$" − $\frac{5}{8}$"	_____
50. $\frac{9}{16}$" + $\frac{1}{4}$"	_____	70. 3$\frac{5}{8}$" − $\frac{7}{16}$"	_____
51. $\frac{3}{4}$" + $\frac{3}{16}$"	_____	71. 1$\frac{1}{4}$" − $\frac{5}{8}$"	_____
52. $\frac{11}{16}$" + $\frac{1}{8}$"	_____	72. 2$\frac{1}{2}$" − $\frac{11}{16}$"	_____
53. $\frac{1}{8}$" + $\frac{3}{16}$"	_____	73. 3$\frac{7}{8}$" − $\frac{15}{16}$"	_____
54. $\frac{3}{8}$" + 1$\frac{1}{4}$"	_____	74. 1$\frac{3}{4}$" + $\frac{5}{8}$" − $\frac{3}{8}$"	_____
55. $\frac{7}{16}$" + 1$\frac{3}{4}$"	_____	75. 1$\frac{1}{2}$" + 1" − $\frac{5}{16}$"	_____
56. 1$\frac{1}{2}$" + $\frac{3}{16}$"	_____	76. 2" − $\frac{7}{16}$" + $\frac{3}{8}$"	_____
57. $\frac{9}{16}$" + 1$\frac{5}{8}$" + $\frac{1}{4}$"	_____	77. 3$\frac{5}{8}$" + $\frac{7}{8}$" − $\frac{1}{4}$"	_____
58. 2$\frac{3}{8}$" + 1$\frac{5}{16}$" + $\frac{3}{4}$"	_____	78. 2$\frac{1}{4}$" − $\frac{3}{4}$" + $\frac{5}{16}$"	_____
59. 3$\frac{1}{8}$" + $\frac{1}{2}$" + $\frac{7}{16}$"	_____	79. $\frac{13}{16}$" − $\frac{3}{8}$" + 1$\frac{5}{8}$"	_____
60. 1$\frac{11}{16}$" + $\frac{5}{8}$" + 2	_____	80. 1$\frac{1}{16}$" + $\frac{7}{16}$" − $\frac{1}{2}$"	_____
61. $\frac{15}{16}$" + 1$\frac{1}{4}$" + $\frac{1}{8}$"	_____	81. 2$\frac{7}{8}$" − 1$\frac{5}{16}$" + $\frac{1}{4}$"	_____
62. $\frac{5}{8}$" + $\frac{13}{16}$" + 2$\frac{1}{4}$"	_____	82. 3" + 2$\frac{1}{4}$" − $\frac{7}{8}$"	_____
63. $\frac{13}{16}$" + 3$\frac{5}{8}$" + $\frac{1}{2}$"	_____	83. 2$\frac{7}{16}$" + $\frac{3}{8}$" − $\frac{5}{16}$"	_____

Unit 15 MULTIPLE OPERATIONS OF COMMON AND DECIMAL FRACTIONS

BASIC PRINCIPLES OF COMMON AND DECIMAL FRACTIONS

The automotive technician has to work with both fractions and decimals. Sometimes the technician has to use both in the same problem. When this happens, convert all fractions to decimals or all decimals to fractions before attempting to solve the problem.

To change a fraction to a decimal, divide the numerator by the denominator.

Example: $\dfrac{11}{16}$

$$
\begin{array}{r}
.6875 \\
16\overline{)11.0000} \\
9\,6 \\
\hline
1\,40 \\
1\,28 \\
\hline
120 \\
112 \\
\hline
80 \\
80 \\
\hline
\end{array}
$$

Answer: .6875

To change a decimal to a fraction, write the number to the right of the decimal point as the numerator. The denominator is a one followed by the same number of zeros as there are numbers to the right of the decimal point.

Reduce the fraction to lowest terms.

Examples: $.5 = \dfrac{5}{10} = \dfrac{1}{2}$

$.375 = \dfrac{375}{1000} = \dfrac{75}{200} = \dfrac{3}{8}$

$.6875 = \dfrac{6875}{10000} = \dfrac{1375}{2000} = \dfrac{275}{400} = \dfrac{11}{16}$

Review Units 5–14.

Review denominate numbers in Section I of the Appendix.

Apply these principles to the automotive field by solving the practical problems that follow.

PRACTICAL PROBLEMS

1. A truck driver travels a distance of 2,026¾ miles in 5½ days. What is the average distance traveled each day? _____

2. A garage owner buys 35,480 gallons of gasoline at $1.588 per gallon. If the gasoline is sold for $1.839 per gallon, what is the profit? _____

3. The minimum width for repair stalls in a garage is 8¾ feet. How many stalls can be built in an area 38 feet wide? _____

4. One gallon of gasoline weights 6.56 pounds. What is the weight of the gasoline in a tank that holds 21.8 gallons? _____

5. A motor with a standard bore of 3⅞ inches is rebored to 0.050 inch oversize. Express the new bore measure as a decimal. _____

6. Six piston pins, 0.005 inch oversize, are required for a certain job. The standard size is ¹¹⁄₁₆ inch. What do the new pins measure? _____

7. The cylinder shown is 3½ inches in diameter and is rebored 0.030 inch oversize. What does the finished diameter read on the micrometer? _____

8. In "feeling" the ring fit in the piston ring groove shown, a 0.002-inch feeler fits snugly. What is the width of the groove if a ³⁄16-inch ring is being used?   _____

9. The holes in the frame holding the side member to the crossmember have become worn and enlarged. Originally the measurements were ³⁄8 inch. The measurement is now about 0.450 inch in diameter.

    a. Express as a decimal the difference between the original and present diameter of the hole.   a. _____

    b. Find, to the nearest ¹⁄16 inch, the difference between the diameters.   b. _____

10. A piston pin measures ¹⁵⁄16 inch in diameter. The clearance for a fit in the connecting rod bushing is 0.0005 inch. What is the hone setting to the nearest 0.0001 inch?   _____

11. What fractional size bolt is used in a 0.314-inch hole? Express the answer to the nearest ¹⁄16 inch.   _____

12. A car travels 1,296¹⁄2 miles and uses 55³⁄4 gallons of gasoline. What is the mileage to the nearest hundredth?   _____

13. What is the cost of 75³⁄4 gallons of gasoline at $1.359 per gallon? Express the answer to the nearest cent.   _____

14. Mr. Smith kept this record of his car expenses for one year. What is the total
cost of operating his car? _____

Item	Amount Used	Cost (per unit)	Cost (each item)
Gasoline	800 gallons	$1.29 per gal. (avg.)	
Transmission Fluid	1 quart	$1.75 per quart	
Power Steering Belt	1 belt	$4.90 each	
Lube Job	6 lube jobs	$5.50 each	
Differential Lube	2 pints	$1.75 per pint	
Oil Change Note: 5 qt. with filter change 4 qt. without filter change	12 oil changes _____ quarts _____ quarts	$1.39 per quart	
Oil Filter every second oil change	_____ filters	$5.40	
Tire Wear		$55.00	
Miscellaneous Repair Charges		$189.40	
Note: Depreciation not considered.		Total Cost	

15. Mrs. Smith's car is valued at $11,990.00, and she allows one-fourth of the
value for one year's depreciation. The expenses for her car are $1,596.00.
What is the operating cost per mile, including expenses and depreciation, if
the car is driven 14,954 miles in one year? _____

16. What is the cost of 350 pounds of pressure grease at $1.87 per pound? _____

17. A grease gun holds 34 pounds of lubricant. How many times can the gun be
filled from a 120-pound drum of lubricant? Express the answer to the nearest
hundredth. _____

18. A transmission holds $4\frac{1}{2}$ pints of oil. One pint of gear oil weighs $1\frac{1}{16}$ pounds.
How many complete transmissions can be filled from a barrel containing 377
pounds of gear oil? _____

19. What is the total cost of these articles: 4 large grease guns, $23.17 each;
6 small grease guns, $9.60 each; 100 lubricator fittings (straight), $0.32 each;
75 lubricator fittings (120 degrees), $0.83 each; 16 lubricator fittings (90
degrees), $0.82 each? _____

20. Find the total cost of these lubricating supplies: one 120-pound drum gear grease, $1.86 a pound; one 35-pound drum of 930-AA high-temperature grease, $1.98 a pound; one 5-gallon can oil, $4.84 a gallon. _____

21. Lubricating supplies cost $678.35. A refund for the return of the empty containers is: $10.12 for the 250-pound drum; $5.80 for the 95-pound drum. What is the net cost of the supplies? _____

22. A customer pays $5.50 for having a car lubricated. The cost of the materials is $1.75, and the mechanic receives $2.50. What amount is left to cover the owner's overhead and profit? _____

23. A mechanic receives $2.50 for greasing a car. How many cars does the mechanic have to grease to earn $32.50? _____

24. What is the micrometer reading for a piece of $7/8$-inch diameter stock? _____

25. What is the micrometer reading of a wrist pin that measures $13/16$ inch? _____

26. A piece of round stock is $5/8$ inch. What is the micrometer reading? _____

27. An automobile has a wrist pin $15/16$ inch in diameter. Find the micrometer reading to the nearest thousandth. _____

28. A certain shaft size is given as $55/64$ inch. What does this measure on a micrometer, to the nearest thousandth inch? _____

29. Express the measurement of a $3/4$-inch wrist pin to the nearest 0.001 inch. _____

30. A wrist pin is $7/8$ inch in diameter. What is the micrometer reading for the diameter of this pin? _____

31. A certain wrist pin measures 0.8594 inch in diameter. What is the fractional dimension to the nearest $1/64$ inch? _____

32. The micrometer reading for the diameter of the wrist pin in the sketch is 0.875 inch. What is this dimension, to the nearest $\frac{1}{32}$ inch?   _____

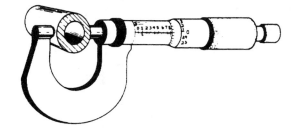

33. Express $\frac{1}{16}$ inch to the nearest 0.0001 inch.   _____

34. A certain stock measures 0.875 inch in diameter. What does this measure, to the nearest $\frac{1}{16}$ inch, with an outside caliper and a scale?   _____

35. What does a piece of $\frac{27}{32}$-inch bar stock measure to the nearest 0.001 inch with a micrometer?   _____

36. To what reading is a micrometer set for $\frac{51}{64}$ inch? Round the answer to the nearest thousandth inch.   _____

37. The diameter of a certain wrist pin is $\frac{49}{64}$ inch. Express this diameter to the nearest 0.001 inch.   _____

38. Express $\frac{31}{32}$ inch to the nearest 0.001 inch.   _____

39. Give the micrometer reading for a piece of $\frac{3}{64}$-inch sheet metal to the nearest 0.001 inch.   _____

40. How many 0.001" long line segments are in $\frac{1}{64}$"?   _____

Note: Use this table for problems 41 through 45.

TABLE OF TAP DRILL SIZES						
National Coarse			National Fine			
Thread Size	Diam. of Hole in Inches	Drill	Thread Size		Diam. of Hole in Inches	Drill
1/4 −20	0.201	#7	1/4	−28	0.213	#3
5/16 −18	0.257	F	5/16	−24	0.272	I
3/8 −16	0.313	5/16	3/8	−24	0.332	Q
7/16 −14	0.368	U	7/16	−20	0.391	25/64
1/2 −13	0.422	27/64	1/2	−20	0.453	29/64
9/16 −12	0.484	31/64	9/16	−18	0.516	33/64
5/8 −11	0.531	17/32	5/8	−18	0.578	37/64
11/16 −11	0.594	19/32	11/16	−16	0.625	5/8
3/4 −10	0.656	21/32	3/4	−16	0.688	11/16
13/16 −10	0.719	23/32	7/8	−14	0.813	13/16
7/8 −9	0.766	49/64	1	−14	0.938	15/16
15/16 −9	0.828	53/64				
1 −8	0.875	7/8				

41. Threads are stripped in a hole that takes a ¼"–20 NC cap screw, and the hole is drilled and tapped for the next larger size.

   a. What diameter hole is drilled?   a. _____

   b. What number drill is used?   b. _____

Note: Use this diagram for problems 42 through 45.

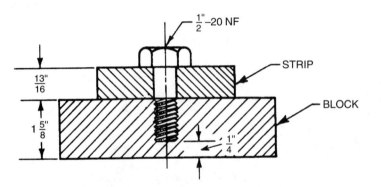

42. It is necessary to drill through the strip into the block without going closer than ¼ inch to the bottom of the block. To what depth is the hole drilled?   _____

43. What diameter hole is drilled?   _____

44. What number drill is used for the hole in the block?   _____

45. A ⅝"–18 NF cap screw must pass through the strip into a tapped hole in the block. What number drill is used for the hole that is to be tapped?   _____

Automotive electrical cable is made of stranded wires for flexibility. The number of strands used varies but is usually 7, 19, 37, 61, 91, or 127. In ordering cable, the diameter is expressed as a gauge size. This table gives the gauge sizes. **Note:** Use this table for problems 46 through 51.

Wire Diameter in Inches	American Wire Gauge	Circular Mil Area	Wire Diameter in Inches	American Wire Gauge	Circular Mil Area
0.4600	0000	211,600	0.0284	21	810.1
0.4096	000	167,800	0.0253	22	642.4
0.3648	00	133,100	0.0225	23	509.5
0.3249	0	105,500	0.0201	24	404.0
0.2893	1	83,690	0.0179	25	320.4
0.2576	2	66,370	0.0159	26	254.1
0.2294	3	52,640	0.0142	27	201.5
0.2043	4	41,740	0.0126	28	159.8
0.1620	6	26,250	0.0112	29	126.7
0.1285	8	16,510	0.0100	30	100.5
0.1019	10	10,380	0.0089	31	79.70
0.0808	12	6,530	0.0079	32	63.21
0.0640	14	4,107	0.0070	33	50.13
0.0508	16	2,583	0.0063	34	39.75
0.0403	18	1,624	0.0056	35	31.52
0.0319	20	1,002	0.0050	36	25.0

Note: The *circular mil* is a unit of area used in electrical work. The values under the column headed *Circular Mil Area* in the chart give a comparison of the amount of copper in each wire.

46. A cable has 19 strands of wire, each strand measuring 0.0112" in diameter (*#29*-gauge). What size cable is ordered to replace this cable? Give the answer as the nearest gauge number.   _____

47.   What is the nearest gauge number for a cable of seven strands of *#18* wire?   _____

48.   Of what gauge single strands does a 37-stranded motor cable *#1*-gauge consist?   _____

49.   Approximately how many times more copper is there in a *#1*-gauge wire than in a *#4*-gauge wire?   _____

50.   What gauge is a cable that has 61 strands of 0.010-inch diameter wire?   _____

51.   The diameter of a *#36* wire is half the diameter of a *#30* wire. How many times more copper does the *#30* wire have?   _____

Percent and Percentage

Unit 16 SIMPLE PERCENT

BASIC PRINCIPLES OF SIMPLE PERCENT

Percent is a way of saying "how many parts per hundred." For example, if five parts out of a hundred were found defective, we could say that 5% were defective.

To change a percent to a fraction, simply place the percent over 100 and reduce the fraction to its lowest terms.

Example: $30\% = \dfrac{30}{100} = \dfrac{3}{10} = 10\overline{)3.00}^{.30}$

To change a percent to a decimal, move the decimal point *two* places to the left.

Example: $1\% = .01$ and $14\% = .14$

If the percent has a fraction in it, as in $1\frac{1}{2}\%$, change the fraction to a decimal first. In this case, it would be 1.5%. Then move the decimal point two places to the left: $.01.5$

To change a decimal to a percent, move the decimal two places to the right: $.333 = 33.3\%$

To find what percent one number is of another, change the numbers to a fraction or a decimal and reduce the result to its lowest terms.

Example: What percent of 80 is 10?

$$\frac{10}{80} = \frac{1}{8}$$

$$8\overline{)1.000}^{.125}$$
$$\underline{8}$$
$$20$$
$$\underline{16}$$
$$40$$
$$\underline{40}$$

$$.125 = 12.5\%$$

PRACTICAL PROBLEMS

Using the following percent, give the fractional equivalent.

1. 25% _____   9. 5% _____
2. 40% _____   10. $8\frac{1}{3}$% _____
3. 10% _____   11. 125% _____
4. 75% _____   12. 500% _____
5. 60% _____   13. 95% _____
6. $12\frac{1}{2}$% _____   14. $33\frac{1}{3}$% _____
7. $66\frac{2}{3}$% _____   15. 2% _____
8. $16\frac{2}{3}$% _____   16. 240% _____

Using the following percent, give the decimal equivalent.

17. 67% _____   27. 500% _____
18. 31% _____   28. 95% _____
19. $1\frac{3}{4}$% _____   29. 48% _____
20. $66\frac{2}{3}$% _____   30. $10\frac{1}{2}$% _____
21. $12\frac{1}{2}$% _____   31. 17% _____
22. $7\frac{1}{3}$% _____   32. 330% _____
23. $43\frac{1}{2}$% _____   33. 125% _____
24. $\frac{5}{8}$% _____   34. $\frac{1}{2}$% _____
25. 87% _____   35. 3% _____
26. $6\frac{1}{3}$% _____   36. $41\frac{2}{3}$% _____

In the following problems, find what percent of the first number is the second number.

Example: What percent of 50 is 20? Divide 20 by 50 = .40 = 40%.

37. Of 64 is 16? _____   42. Of 27 is 9? _____
38. Of 25 is 5? _____   43. Of 99 is 23? _____
39. Of 150 is 15? _____   44. Of 85 is 5? _____
40. Of 160 is 20? _____   45. Of 32 is 8? _____
41. Of 120 is 6? _____   46. Of 63 is 9? _____

47. Of 234 is 50?   _____

48. Of 64 is 12?   _____

49. Of 12 is 5?   _____

50. Of 3 is 1?   _____

51. Of 412 is 13?   _____

52. Of 978 is 651?   _____

53. Of 14 is 3?   _____

54. Of 24 is 8?   _____

55. Of 77 is 7?   _____

56. Of 10 is 4?   _____

57. Of 136 is 25?   _____

58. Of 16 is 5?   _____

59. Of 38 is 10?   _____

60. Of 716 is 36?   _____

This table shows the percent ethylene glycol and the corresponding protection levels. The percent of ethylene glycol is found by using the formula,

$$\text{Percent ethylene glycol} = \frac{\text{gallons of ethylene glycol}}{\text{gallon capacity}}$$

Note: Use this chart for problems 61–74.

% Ethylene glycol	Protection to
20	+16°F
25	+10°F
30	+ 4°F
33 1/3	0°F
40	−12°F
50	−34°F
60	−62°F

Using the table, find the protection levels for problems 61 through 74 in the following chart.

Hint: First find the percent of Prestone, then read the corresponding protection level.

PROTECTION LEVEL CHART

Gallon Capacity	Gallons of Ethylene Glycol Used				
	1	1 1/2	2	2 1/2	3
2 1/2	61	62			
3	63	64			
4	65		66		
5	67	68	69	70	71
6		72	73		74

61. _____ %   _____ °F   68. _____ %   _____ °F

62. _____ %   _____ °F   69. _____ %   _____ °F

63. _____ %   _____ °F   70. _____ %   _____ °F

64. _____ %   _____ °F   71. _____ %   _____ °F

65. _____ %   _____ °F   72. _____ %   _____ °F

66. _____ %   _____ °F   73. _____ %   _____ °F

67. _____ %   _____ °F   74. _____ %   _____ °F

75. If 15 out of 75 piston pins are rejected, what percent is rejected?   _____

76. If 15 out of 75 piston pins are rejected, what percent is accepted?   _____

77. A repair shop reduces the working hours from 40 to 35 without reducing the weekly rate of pay of the mechanics. What is the percent of decrease in hours?   _____

78. Six quarts of antifreeze are used in the cooling system of a truck that holds 5½ gallons. What percent of the solution is antifreeze? Express the answer to the nearest tenth percent.   _____

79. A small car equipped with a diesel engine gets 56 miles per gallon (mpg). The same car equipped with a gasoline engine gets only 41 mpg. What is the percent increase in mileage of the diesel over the gasoline engine?   _____

80. An owner finds that when he uses a clamshell luggage carrier on top of his car, it cuts his gas mileage 2.3%. If he normally gets 34 mpg, what is his mileage when he uses the carrier?   _____

81. Engineers find that a small diesel engine develops 55 brake horsepower (bhp). By adding a turbocharger, the horsepower increases to 68 bhp. What is the percent increase in horsepower?   _____

82. A survey finds that 128 fleets in the United States and Canada order 93,032 cars this year and that 70% are compact and subcompact cars. How many full-sized cars are ordered?   _____

83. If 4 out of every 10 car sales are made directly to women, what percent of car sales are made to women?   _____

84. An experimental car gets 65 mpg when driven at a steady 35 miles per hour (mph). At 55 mph, the mileage drops to 49 mpg. What is the percent decrease in gas mileage?   _____

Unit 17 SIMPLE PERCENTAGE

BASIC PRINCIPLES OF PERCENTAGE

Percentage is a part of a whole number expressed in hundredths. There are either three or four steps in finding a percentage, depending on whether there are one or two decimal points in the problem.

Example: Here is a single-decimal-point problem. There are 80 cars in the parking lot, and 15% have some red paint. How many cars have red paint?

Step 1. Change the percentage into a decimal fraction by moving the decimal point 2 places to the left.

15% is written as .15

Reminder: All whole numbers have a decimal point after the last number on the right even though we do not put it in: 26 is 26., 39 is 39., 426 is 426.

Step 2. Multiply the number of cars by .15.

$$
\begin{array}{r}
80 \\
.15 \\
\hline
400 \\
80 \\
\hline
1200
\end{array}
$$

Step 3. Count the digits to the right of the decimal point. There are 2 digits. Go to the answer 1200. Recall that the decimal point is after the last zero. Move it 2 places to the left:

12.0.0. or 12.00

Answer: There are 12 cars with red paint.

Example: Here is a problem with two decimal points. Six spark plugs sell for $12.60. The dealer took off 25%. How much did I save?

You do steps 1–3 again. The only difference is that you will have to count the digits after both decimal points in step 3.

Step 1. Change the percentage into a decimal fraction:

25% is written as .25

Step 2.   Multiply

$$
\begin{array}{r}
\$12.60 \\
.25 \\
\hline
63\ 00 \\
252\ 0 \\
\hline
\$315\ 00
\end{array}
$$

Step 3.   Count the total digits to the right of *both* decimal points. There are 4 places.

Start at the end of $31500. and count 4 places to the left.

$31500. $3.1 5 0 0 or $3.1500

Answer: My discount was $3.15.

There are two ways to solve the following problems. Use the formula:

$$\text{Percentage} = \text{base} \times \text{rate}$$

Base is the number that you want to find the percentage of.

Rate is the percent changed to a decimal by moving the decimal point two places to the left.

Example: <u>Rate</u>   <u>Base</u>   <u>Percentage</u>
             3%   of   1,500 parts =   percentage
             .03   ×   1,500 parts =   45 parts

One easy way to use the percentage formula is this triangle:

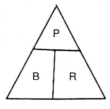

To find the percentage, place your finger over the P.   P = B × R.

To find the rate, place your finger over the R.   R = P ÷ B.

To find the base, place your finger over the B.   B = P ÷ R.

Another way to help solve the following problems is to use the percent circle:

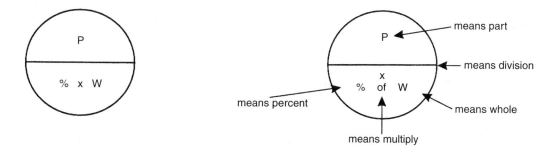

Example: Find 3% of 1,500.

To use the circle, determine what you are given in the problem. In this case, you are given percent (3%), the key word "of," and 1,500. Fit the information into the circle, with 3% replacing the % symbol, the times sign replacing the word "of," and 1,500 replacing the "W." Notice how the original problem fits into the bottom half of the circle. Next, change the 3% into a decimal: .03 × 1,500 = 45.

Example: What percent of 80 is 10?

Step 1.   Determine what you are looking for and what information you are given in the problem. In this case, you are looking for percent itself. Therefore, on the circle, the other numbers represent either "P" or "W."

Step 2.   To determine which number is which, notice the key word "of," next to the number 80. "Of 80" means that 80 goes in the bottom half of the circle, and therefore, 10 is "P." The problem should be set up as follows: $^{10}/_{80} = \frac{1}{8}$.

Remember, the problem asks for the percent, so the answer $\frac{1}{8}$ must be changed into percent by

dividing: $8\overline{)1.000}$ .125

The decimal .125 must be changed into a percent by moving the decimal point two places to the right and placing a percent sign next to the number: 12.5%.

Example: 12.5% of a number is 10. What is the number?

Step 1.   Determine what you are looking for. In this case, it is "a number." On the percent circle, the answer must be either "P" or "W."

Step 2:   Look for key words in the problem. The 12.5% fits in the % on the bottom half of the circle. "Of a number" also belongs on the bottom half. The 10 can only belong on top as "P." What we have, therefore, is "P" divided by %, or 10 ÷ .125 (don't forget to change 12.5% into a decimal first). The number is 80.

When solving word problems, follow the same steps.

Example: A mechanic's helper earning $6.40 per hour receives a 10% increase in pay. What is the new hourly rate?

Step 1.   What does the problem ask for, and what do we know? The problem asks for the *new hourly rate*. We know the % increase (10%), and we know the whole or total amount before the increase ($6.40). Therefore, we are looking for "P."

Step 2.   Since we are looking for "P," we know from the circle that we must multiply % (in its decimal form, .10) by "W" (6.40), which equals .64. Note: This is *not* the new hourly rate but the percent increase. To find the new hourly rate, add the .64 to $6.40, which equals 7.04. The new rate is $7.04 per hour.

PRACTICAL PROBLEMS

1.   If a mechanic's hourly rate of $10.15 is increased $16\frac{2}{3}$%, what is the new hourly rate to the nearest cent?   _____

2.   A mechanic who is paid $488.00 per week gets an 8% increase. How much more per week does the mechanic get?   _____

3.   A repair shop gives a 5% bonus to mechanics finishing jobs in a given amount of time. The mechanics work 40 hours a week at an hourly rate of $14.30. How much more does a mechanic receive if the bonus is earned?   _____

4.   The charge for labor in having the front end of a car rebuilt is $27.50 per hour. The technician gets 53% of the total amount. If the job takes 4 hours, how much does the technician get?   _____

5.   A mechanic's helper receives $5.35 per hour. A raise in pay of $12\frac{1}{2}$% is given. What is the new hourly rate of pay?   _____

6.   A shipment of one gross (144) automobile headlight lamps is dropped during delivery. The number of broken lamps is 50% of the shipment. How many lamps are broken?   _____

7.   A garage owner receives a shipment of 200 taillights. The owner finds that 25% of the shipment is damaged. How many taillights are damaged?   _____

8. In a shipment of 500 side-view mirrors, 10% arrive cracked. Find the number of cracked mirrors. _____

9. A mechanic orders 80 cans of motor oil. The number of dented cans is 12½% of the shipment. How many cans are dented? _____

10. In a shipment of 120 lamps, 16⅔% arrive broken. How many lamps are broken? _____

Note: Use this information for problems 11–13.

1) *Babbitt metal* for severe service is made of about 90% tin, 5% copper, and 5% antimony.
2) *Babbitt metal* is the soft surface on connecting rods and main bearings in the engine.
3) *Severe service* is high speeds at high temperatures and loads.

11. If 25 pounds of babbitt metal are needed, how much copper is used? _____

12. Ten pounds of antimony is used in how many pounds of babbitt metal? _____

13. How much tin is needed for 42 pounds of babbitt metal? _____

14. One percent of the supply of gasoline is lost per day through evaporation. How many gallons are lost each day from 4 tanks each holding 2,500 gallons? _____

15. A company sells 300 tires, of which 68% are radials. How many radial tires are sold? _____

16. Suppose 162,000 mechanics are certified by Automotive Service Excellence Tests. If 18.6% of these pass seven more tests to become Certified Master Automotive Technicians, how many mechanics will have passed all eight tests? _____

17. The P195/75R14 is the most popular radial tire sold, at 17% of all radials. If a dealer sells 500 radials a year, how many P195/75R14s should the dealer expect to sell? _____

18. A garage completes 538 jobs, with a comeback rate of 1.29%. How many jobs do they have to do over again? _____

19.   National figures show that 82.3% of new car buyers order air conditioning. If a dealer expects to order 356 cars, how many should the dealer order with air conditioning?   _____

20.   A car owner receives notice that his car insurance will be increased 3.2%. He now pays $173.60. How much will he have to pay next time?   _____

21.   A mechanic earns $567.90 in a week. The FICA tax is 8.5%. How much tax does the mechanic have to pay?   _____

22.   If 15% of a liter of oil is additives, how many cubic centimeters of this liter is lubricating oil?   _____

23.   A dealer finds that 23.2% of her new car sales become regular service customers after their warranty period is over. If she sells 1,253 cars in a year, how many service customers will she gain?   _____

24.   A tune-up specialist finds that his business increases 9.7% after advertising on the radio. If he does 41 tune-ups a week before advertising, how many more should he average now?   _____

25.   The cost of doing business for a garage went up 7.4% in one year. If the owner spent $17,565.00 last year, what will the owner spend this year?   _____

26.   Goodyear sells 14.5% of the replacement passenger car tires. If 1,000,000 tires are sold in a region, how many does Goodyear sell?   _____

27.   The next highest in tire sales is Sears at 9%. How many tires will Sears have sold in the same region?   _____

28.   By turbocharging a diesel engine, a mechanic increases the brake horsepower (bhp) 12.4%. The improved engine is rated at 68 bhp. What was the original engine producing?   _____

29.   A tool salesperson notes that sales of metric socket sets increase from 23 to 27 sets per month. What is the percentage of increase?   _____

    (**Hint:** Percentage of increase $= \dfrac{\text{amount of increase}}{\text{original amount}}$)

30. In a state emission inspection program, about 26% of the diesel vehicles fail the first inspection. If one station inspects 135 diesels in one month, how many will fail the test?   _____

31. In the same state, the fail rate for gas engines is only 18%. If a station inspects 355 vehicles in one month, how many will fail?   _____

32. If there are 40 problems in this unit and a student answers 37 of them correctly, what percent does the student get correct?   _____

33. If 0.48% of 2,500 cars inspected fail for visible smoke at the tailpipe, how many cars fail?   _____

34. The EPA Tampering Survey reports that cars with emission controls such as the catalytic converter removed, emit as much as 500% more emissions. If a car with a catalytic converter tests at 0.5% carbon monoxide (CO), what will it test with the converter removed?   _____ % CO

 *Unit 18* **DISCOUNTS**

BASIC PRINCIPLES OF DISCOUNTS

A discount is a reduction in the list price of a part to encourage sales. Change the percent of the discount to a decimal. Multiply the list price by the discount, and count off the correct number of decimal places in the answer.

Example: Find 28% off the list price of $32.50.

$$\begin{array}{r} \$32.50 \\ \times\ \ .28 \\ \hline 260\ 00 \\ 650\ 0\ \ \\ \hline \$9.10\ 00 \end{array}$$

$$\begin{array}{r} \$32.50 \text{ list price} \\ -\ 9.10 \text{ discount} \\ \hline \$23.40 \text{ selling price} \end{array}$$

PRACTICAL PROBLEMS

1. A mechanic purchases 7½ feet of ¹³⁄₃₂-inch air-conditioning hose at $4.18 per foot. A discount of 40% is given. What does the mechanic pay for this hose to the nearest cent?   _____

2. A rear-seat speaker switch costs $4.90. If the discount is 33⅓% what is the net cost of the switch to the nearest cent?   _____

3. Three-eighths inch compression fittings cost $2.07 each in lots of six. What is the cost of 6 fittings if the discount is 33⅓?   _____

4. The list price of ½-inch flare nuts is 95¢ each. If the discount is 25%, what will 12 nuts cost?   _____

5. Number 1157 light bulbs cost 65¢ each, list price. The discount on a purchase of 12 is 25%. What is the net cost of 12 light bulbs?   _____

6. A jobber sells power steering pressure hose to a garage owner at a discount of 40%. If the list price is $6.31 per foot, what is the garage owner's cost per foot? Express the answer to the nearest cent. _____

7. In overhauling an engine, a mechanic uses 1 gasket set, $61.65; 1 set of piston rings, $60.80; 6 main bearings, $7.25 each; 6 rod bearings, $4.95 each; 1 main thrust bearing, $13.20; 2 valves, $9.20 each; 1 timing chain, $25.45. The parts store gives a 36% discount. What is the mechanic's profit on the parts to the nearest cent? _____

8. A mechanic needs two $\frac{3}{8}$" bolts that cost 19¢ each and two nuts that cost 9¢ each. If a discount of $33\frac{1}{3}$% is allowed, what is the cost of the parts? Express the answer to the nearest cent. _____

9. The cost of the parts to repair a car is $96.70. A $33\frac{1}{3}$% discount from the list price is given on parts. What is the net cost of the parts? _____

10. A 40% discount is given on *#12*-gauge wire. When purchased in 500-foot rolls, an additional 10% discount is given. If the list price is 17¢ per foot, what is the net cost of a 500-foot roll? _____

11. The monthly accounts owed by a service station total $1,467.62. If these accounts are paid on or before the 10th of the next month, a 2% discount is given. How much does the service station owe after the discount is taken? Express the answer to the nearest whole cent. _____

12. A vacuum pump that has been selling for $25.50 is put on sale for $14.98. What percent discount is this? _____

13. A tool salesperson puts all tools on sale at 5.5% off list price for cash. How much can a mechanic save on tools that would have sold at a list price of $1,153.34? _____

14. A new car buyer receives a rebate direct from the manufacturer of $1,400.00 on a car that costs $16,235.00. How much of a discount is this? _____

15. On sale, a remanufactured engine is offered at 8% discount. The regular price for this engine is $1,963.00. What is the savings by buying it on sale? _____

16. A discount store offers four shock absorbers for the price of three. What is the rate of discount? _____

17. A customer purchases 2 gallons of antifreeze at $4.29 a gallon, with a mail-in rebate of $1.00 a gallon. What is the percent of the rebate?   _____

18. If a garage owner purchases 10 mufflers at a time from a jobber, the owner receives an additional discount of 12%. The regular price is $22.35 each. How much does the owner save by purchasing 10 at a time?   _____

19. A manufacturer of testing equipment offers a scanner for testing computer-controlled cars for the introductory price of $327.50 if purchased before the end of the month. After that, the price will be $385.00. How much of a discount is this?   _____

20. A parts store allows a discount of 2% on monthly charges paid before the 10th of the following month. How much can a mechanic save on a bill of $454.83 by paying it before the 10th?   _____

21. A jobber offers a discount of 4% to its best customers in addition to their discount of 35%. What is the cost of parts that list at $1,054.78 with these discounts? (**Hint:** First take the regular discount from the bill, and then take the second discount on the remaining amount.)   _____

22. A transmission rebuilder offers a second discount of 7% to its customers that purchase over 4 transmissions in 1 month. The regular discount on a transmission that lists for $535.00 is 24%. What does the shop owner have to pay if she purchases 6 transmissions in a month?   _____

23. A tune-up specialist installs the following parts with these prices:

    Electronic ignition module @ $65.98
    EGR valve @ $55.95
    PCV valve @ $4.95
    8 spark plugs @ $2.97 each
    Fuel filter @ $3.57
    Air filter @ $7.73

    He receives a 31% discount. How much does he have to pay for the parts?   _____

24. A transmission specialist overhauls an automatic transmission and uses the following parts:

Master overhaul kit @ $111.70
Front pump @ $52.80
Rebuilt torque converter @ $65.00
9 quarts of transmission fluid @ $1.65 a quart

If the specialist gets a 27% discount on transmission parts, how much is made on the parts?   _____

25. An oil company offers a $3.00 rebate on the purchase of a case of motor oil (24 quarts) that sells for 94¢ a quart. If the buyer sends for the rebate, what is the percent of discount?   _____

 # Unit 19 PROFIT AND LOSS, COMMISSIONS

BASIC PRINCIPLES OF PROFIT AND LOSS, COMMISSIONS

Profit is the amount left over after all expenses are paid. If the expenses are more than the selling price, the difference is a *loss.* A *commission* is a percentage of the selling price paid to the salesperson. Percentage is frequently used to express profit, loss, or commission.

Example: $5,120.00 (selling price of a used car)
        − 4,089.13 (trade-in price, repairs, and salesperson's commission)
        $1,030.87 (profit)

The percentage of profit (or margin of profit) is found by dividing the profit by the selling price.

Example: $\dfrac{1,030.87}{5,120.00}$

$$\begin{array}{r} 0.20134 \\ 5,120 \overline{)\, 1,030.87} \\ 1\ 024\ 0 \\ \hline 6\ 870 \\ 5\ 120 \\ \hline 1\ 7503 \\ 1\ 5360 \\ \hline 21430 \\ 20480 \\ \hline \end{array}$$

        0.20134 = 20.134%

PRACTICAL PROBLEMS

1. The total receipts of a repair shop for one month are $6,000.00. Of this amount, 40% is for wages, 10% is for rent, 5% is for heat, and 20% is for other overhead expenses. Does the shop show a profit?   _____

    If yes, what is the profit expressed as a percent?   _____

    What is the profit expressed in dollars and cents?   _____

2. A garage owner buys a job lot of 500 feet of *#12*-gauge wire. The price is $0.25 per foot. The next day, the price of this wire increases $12\frac{1}{2}\%$. How much is saved by purchasing the wire at the lower price? Express the answer to the nearest cent.

    _____

3. At a sale, a garage owner spends $725.45 for a number of different size tires. The freight charges are $26.30. The tires are sold so that a profit of $33\frac{1}{3}\%$ is made. How much must the tires be sold for?

    _____

4. A certain make of automobile priced at $10,342.00 depreciates 23% in the first year. How much money does a purchaser lose in depreciation in one year?

    _____

5. A used car priced at $2,875.00 is reduced 23%. What is the sale price of the car?

    _____

6. A drum of grease contains 54 gallons and weighs 7 pounds per gallon. Through wastage, 5% of the grease is lost. What is the net loss to the garage owner if the grease costs $97\frac{1}{2}$¢ per pound? Express the answer to the nearest cent.

    _____

7. A mechanic's cost chart for flushing out and refilling the rear end of a car is:

   Flushing liquid (2 quarts) @ $1.54 per gallon
   Rear axle oil—for refill (4 pints) @ $2.05 per pint
   Labor charges @ $10.00
   Note: 100% profit on material

   What is the charge to the customer?

    _____

8. An automobile costs $7,527.00 and is sold at a profit of 15% on the cost price. What is the profit?

    _____

9. A mechanic receives a commission of 5% on parts sold to customers. A customer spends $29.00 on parts. What is the mechanic's commission?

    _____

10. A used car salesperson receives $72.50 commission for selling a $1,650.00 car. What is the rate of commission to the nearest hundredth percent?

    _____

11. A car sells for $12,210.00, and a commission of 15% is given. What is the amount of commission paid?

    _____

12. A mechanic buys a new impact wrench that increases the amount of work he can do by 7%. His pay is $650 a week before he buys the wrench. What is his new pay per week?   _____

13. A mechanic, working on the flat-rate system, turns in $1,256.87 labor in one week. If he is on a 45% commission, what is his pay?   _____

14. A service station experiences a drop in business when a new station opens across the street. Business is down 6.4% from $7,436.79. What is the amount of business after the drop?   _____

15. After buying a new oscilloscope, a tune-up shop notes that business increases from $11,547 a week to $12,698. What is the percent of increase?   _____

16. A shop owner offers a mechanic a 3% commission on the first $200 worth of parts sold and 4% on anything over $200. What commission does the mechanic earn on $247 worth of parts sold?   _____

17. A garage notes that the number of repair orders written increases 3% each year from 5,240 per year for two years. How many repair orders is the garage writing now?   _____

18. A change in traffic pattern causes the amount of gasoline sold at a service station to drop 21.7%. If this station made a profit on gasoline last year of $3,126.27, what will the profit be this year?   _____

19. A battery sells for $62.80 list price. The mechanic pays $48.30 to the parts store. What is the mechanic's percent of profit?   _____

20. A mechanic who averages $625 a week labor has a slow week and makes only $475. What is the percent of loss from the mechanic's average pay?   _____

21. A mechanic's profit on brake parts is 45%, and on grease seals 23%. The mechanic does a brake job and uses the following parts:

    A set of disc pads @ $31.50/set
    Rear shoes @ $27.80/set
    Two front grease seals @ $4.73 each

    How much profit does the mechanic make on these parts?   _____

22. An engine overhaul costs the customer $2,135.15. The shop owner figures that $1,110.39 is the cost of labor and parts. What percent profit is made on this job? _____

23. A tune-up shop owner purchases a computerized engine analyzer for $21,300.00. The owner believes that it should pay for itself in five years with the additional profits. How much additional profit must the shop do each year to pay for the machine? _____

24. A shop charges $60.50 an hour for customer labor. Of this amount, the mechanic gets 45%. Of the 55% remaining, the shop owner figures that 40% must take care of overhead costs, leaving 15% for profit. How much profit is in 1 hour of labor? _____

25. A mechanic's commission is increased from 45% to 48% at a flat-rate charge of $60.00 an hour. How much more can the mechanic expect in pay for 49 hours the first week? _____

Unit 20 INTEREST AND TAXES

BASIC PRINCIPLES OF INTEREST AND TAXES

Figuring interest and taxes is another way a technician can use percentage. The amount borrowed is called the *principal.* The amount charged for the use of the principal is the *interest.* The interest rate is usually expressed as a percent. The *term* is the amount of time of the loan. The amount paid equals the principal plus interest.

Example: To buy a four-gas analyzer, a shop must borrow $8,000.00 for 1 year at a rate of 13%. How much will be repaid at the end of the year?

Answer: Change the percent to a decimal fraction: 13% = 0.13.
Multiply the principal by the decimal fraction: $8,000 × 0.13 = $1,040.
Add the principal and interest to get the total: $8,000 + 1,040 = $9,040.

If the term of the loan is more or less than a year, divide the annual interest by 12 months, and then multiply by the number of months.

Example: If the loan is to be repaid in 15 months instead of 1 year, find the total amount repaid.

Answer: The interest for 12 months is $1,040: $1,040 ÷ 12 months = $86.67.
Multiply the interest per month by the number of months: $86.67 × 15 = $1,300.05.
Add the principal and the total interest: $8,000 + $1,300.05 = $9,300.05.

In the two examples, the loan (principal and interest) is paid off at the end of the time period. Most loans are made to be paid monthly or quarterly. The best loan is one that charges *simple interest on the declining balance.* This assesses interest only for the period the money is used. The formula used is as follows:

$$I = (i ÷ n) × B$$

Where: I = interest due
i = annual percentage rate (APR) in decimal form
n = number of payments in a year
B = outstanding loan balance

Example: On the loan of $8,000 at 13% for one year, the interest for the first month is:

$$I = (i \div n) \times B \qquad I = (.13 \div 12) \times \$8,000$$

$$I = \frac{.13}{12} \times \frac{8,000}{1} \qquad I = \frac{260}{3} = \$86.67 \text{ rounded}$$

Lending institutions have these calculations on charts or computer programs. A chart for the loan in the example is illustrated here. If you add the interest paid, you will see that the shop will save a lot more money by borrowing this way.

A loan of $8,000.00 with payments made 12 times a year at interest of 13.00% Each payment $714.54 Requires payment for 1.0 years Last payment will be $714.52			
PERIOD	PRINCIPAL	INTEREST	BALANCE
1	$ 627.87	$ 86.67	$ 7,372.13
2	$ 634.68	$ 79.86	$ 6,737.45
3	$ 641.55	$ 72.99	$ 6,095.90
4	$ 648.50	$ 66.04	$ 5,447.40
5	$ 655.53	$ 59.01	$ 4,791.87
6	$ 662.63	$ 51.91	$ 4,129.25
7	$ 669.81	$ 44.73	$ 3,459.44
8	$ 677.06	$ 37.48	$ 2,782.38
9	$ 684.40	$ 30.14	$ 2,097.98
10	$ 691.81	$ 22.73	$ 1,406.17
11	$ 699.31	$ 15.23	$ 706.86
*12	$ 706.86	$ 7.66	$ 0.00

*Note: Last payment is required at the end of the loan

Tax rates are expressed as percents or in one of these three ways for property:
1.   Mills per dollar value (1 mill equals $\frac{1}{1000}$ of a dollar or $\frac{1}{10}$ of a cent)
2.   Dollars per hundred dollars of value
3.   Dollars per thousand dollars of value

Example: A body shop is assessed at a tax rate of 15 mills on a property valued at $200,000. What is the tax bill?

Answer: If a mill is $\frac{1}{10}$ of a cent: 15 mills = 0.015 cents.

Multiply: $200,000 × 0.015 = $3,000.

PRACTICAL PROBLEMS

1.   How much interest will $300.00 earn in one year if the interest rate is $7\frac{1}{4}$%?   _____

2.   A garage owner's bank account averages $36,000.00 for two years at an interest rate of $5\frac{1}{2}$% per year. How much interest does this account yield?   _____

3.   A garage owner spends $1,000.00 on equipment. The interest charge is 1.5% per month on the unpaid balance. The equipment is paid for in 20 monthly payments of $50.00 each plus interest. This chart shows the payment schedule.

Month	Unpaid Balance	Interest 1.5% of unpaid balance	Amount Paid ($50.00 + Interest)
1	$1,000.00	$15.00	$50.00 + $15.00 = $65.00
2	950.00		
3			
4			
5			
6			
7			
8			
9			
10			
11			
12			
13			
14			
15			
16			
17			
18			
19			
20			
TOTAL		a.	b.

Fill in the chart and find:

a. How much interest is paid.   a.   _____

b. The total amount paid.   b.   _____

c. What percent of the loan the interest is.   c.   _____

4. Some banks make loans at 12% *interest discounted* or *add-on interest.* This means the interest is figured and deducted from the amount being borrowed. For example, if $100.00 is borrowed at 12%, the interest will amount to $12.00. The $12.00 is deducted from the $100.00, and the borrower receives only $88.00 but has to pay $100.00. This means that $12.00 is paid for a loan of $88.00.

   a. What is the actual rate of interest to the nearest tenth percent?   a. _____

   b. Will the rate of interest be the same if $500.00 is borrowed at 12% add-on interest?   b. _____

5. What percent does a tax rate of 20 mills per dollar equal? That is, the tax is what percent of the value?   _____

6. What percent does a tax rate of $3.50 per hundred dollars equal?   _____

7. What percent does a tax rate of $29.50 per thousand dollars equal?   _____

8. An automobile repair shop owner has a plant with equipment valued at $125,000.00. If the tax rate in the community is 25 mills, what is the tax bill?   _____

9. A car is valued at $1,750.00. If the personal property tax rate is $28.00 per thousand, how much is the owner taxed?   _____

In banking, there is a way to tell how long it will take an amount to double in a savings account. It is called the "Rule of 72." Divide 72 by the interest rate to determine the number of years it will take to double the amount.

Example: 72 ÷ 10% = 7.2 years

10. If a technician deposits $1,000.00 in a bank at 5%, how long must the technician wait for the amount to double?   _____

11. If the amount is moved to a credit union that pays 7%, how much time will it take to double?   _____

Another rule is the "Rule of 78," a rule used by car dealers and other lenders. This method charges more interest early in the loan than do other methods. If the loan is repaid early, more interest is paid. If the loan is carried the full term, there is no difference.

This is how it works for a 12-month loan: The sum of the numbers 1 through 12 is 78. The first month, you pay $^{12}/_{78}$ths of the total finance charge. The second month, you pay $^{11}/_{78}$ths, and so on each month until the loan is paid. The rule also works for longer loans. A loan for 24 months is figured this way: The numbers 1 through 24 add up to 300, so the first month's interest is $^{24}/_{300}$ths of the interest charge. The next month's interest is $^{23}/_{300}$ths of the total interest charged, and so on.

Unit 22 PERCENT OF ERROR AND AVERAGES

BASIC PRINCIPLES OF PERCENT OF ERROR AND AVERAGES

Percent of error is the difference between what something does and what it should do.

Example: A thermometer registered 96°F, but the actual temperature was 90°F. Find the difference in the two readings:

$$
\begin{array}{r}
96° \\
-90° \\
\hline
6°
\end{array}
$$

Divide the difference by the true reading:

$$6° \div 90° = .067 \text{ rounded}$$

Change .067 to a percent: 6.7% error.

Averages are found by adding the sum of the units to be averaged and then dividing by the number of units.

Example: A salesperson made four trips of 58 miles, 74 miles, 48 miles, and 64 miles.

$$
\begin{array}{r}
58 \\
74 \\
48 \\
64 \\
\hline
4)\,\overline{244}
\end{array}
\qquad 61 \text{ miles (average)}
$$

PRACTICAL PROBLEMS

1. The speedometer of a car shows 50 mph. The actual speed is 46 mph. What is the percent of error to the nearest tenth?   _____

2. A car is traveling at a speed of 42 mph. The speedometer reading is 35 mph. What is the percent of error to the nearest tenth?   _____

3. The amount of error allowed in the size of structural steel is $\frac{1}{10}$ of 1%. If an
    I beam is 8 inches deep, what, in inches, is the amount of error allowed?   _____

4. Three mechanics in a shop receive $10.85 per hour, $9.05 per hour, and
    $8.70 per hour. What is the average hourly pay?   _____

5. On a trip the miles traveled are: Monday, 328 miles; Tuesday, 461 miles;
    Wednesday, 395 miles; Thursday, 407 miles. What is the average number of
    miles traveled per day?   _____

6. A driver keeps a record of odometer readings and the amount of gas
    purchased. What is the average number of miles per gallon of gasoline?
    Express the answer to the nearest tenth.   _____

Odometer Reading (in miles)		No. of Gallons of Gas Purchased		
Start	28,352	Start	3	in tank
	28,352		7	
	28,480		10	
	28,590		6.4	
	28,700		8.7	
Finish	28,875	Finish	2	in tank

7. On a test, the marks for ten students are: 2 students, 85%; 4 students, 78%;
    3 students, 66%; and 1 student, 92%. What is the average mark on the test?   _____

8. On one tank of gasoline, a car gets 23.5 mpg. The next day it gets 25.8 mpg,
    and the following day 21.5 mpg. What is the average for the three days?   _____

9. Before trading in a car, the car owner wears out 3 sets of tires. On the original
    set, the owner gets 24,320 miles. On the second set, the owner gets 39,141
    miles, and on the third set 35,460 miles. How many miles does the owner
    average per set of tires?   _____

10. Six batteries are tested with the following readings: 12.67 volts, 12.3 volts,
    12.45 volts, 12.01 volts, 12.35 volts, and 12.56 volts. What is the average
    reading?   _____

Measurement

Unit 22 ENGLISH LINEAR MEASUREMENTS

BASIC PRINCIPLES OF ENGLISH LINEAR MEASUREMENT

The following is the table of English Linear Measurement. It is also known as a conversion table.

1 foot	=	12 inches
1 yard	=	36 inches or 3 feet
1 rod	=	16.5 feet or 5.5 yards
1 mile	=	5,280 feet or 1,760 yards or 320 rods

Refer to Unit 24 for more information on fractional measurements of an inch.

Sometimes the hardest part about converting one unit of measurement into another is figuring out whether to multiply or divide. The "box method" can help.

Step 1. Read the problem to determine the *given units* in the problem and the units the answer must be in.

Given units: _____. Ending units: _____.

Step 2. Find the appropriate conversion. The appropriate conversion is the one with the same units in it as in the problem, although the order may be different.

Step 3. If the units in step 1 follow the *same order* (from left to right) as in the conversion, then multiply. If the units in step 1 are *opposite* the order in the conversion, then divide.

Example: An automobile frame measures 9 *feet*. What is the total length in *inches*?

Step 1. Given unit: _feet_. Ending units: _inches_. The order of units from left to right is *feet to inches.*

Step 2. The appropriate conversion is 1 foot = 12 inches.

Step 3. The order of units from left to right in the *problem* follows the same order as in the conversion. This indicates that you multiply 9 and 12, equaling 108 inches.

Example: An automobile frame measures 102 inches. What is its length in feet?

Step 1.   The order of units in the problem is *inches to feet.*

Step 2.   The appropriate conversion is 1 <u>foot</u> = 12 <u>inches</u>.

Step 3.   The order of units in the problem (from left to right) does *not* follow the order of units in the conversion. This tells you to divide 102 by 12. The answer is 8.5 feet.

PRACTICAL PROBLEMS

1.   An automobile frame measures 9 feet 6$\frac{5}{8}$ inches. What is the total length in inches?   _____

2.   A roll of $\frac{3}{8}$-inch tubing is 20 feet long. One piece 6 feet 9$\frac{3}{4}$ inches long is cut from the roll. What length piece remains? Express the answer in feet and inches.   _____

3.   Two pieces of copper tubing, each measuring 3 feet 8$\frac{7}{8}$ inches, are cut from a coil 10 feet long. Find the remaining length in feet and inches.   _____

4.   A roll of ignition wire measures 25 feet. Lengths of 2 feet 3$\frac{7}{16}$ inches, 6 feet 11$\frac{1}{4}$ inches, and 11 feet 9$\frac{3}{8}$ inches are cut from the roll. How many feet and inches of wire remain?   _____

5.   In a repair shop, the owner marks off repair stalls each 8 feet 5$\frac{3}{4}$ inches wide. What is the total distance needed to make six stalls? Express the answer in feet and inches.   _____

6.   The front of a parking lot is 250 feet wide. A space of 7 feet 6 inches is allowed for each car, and 12 feet 9 inches is allowed for a roadway in the center. How many cars can be parked in the front of the lot? (Hint: Express dimensions in feet.)   _____

7.   One mile is equal to 5,280 feet. How many yards are in a mile?   _____

8.   Express 8.65 miles in feet.   _____

9.   Express 6.75 miles in yards.   _____

10.   Express 5 feet 3 inches in yards. Give the answer in fractional form.   _____

11. Express 2 feet 1½ inches in inches. Give the answer in decimal form.   _____

12. Express 8 feet 10¼ inches to the nearest thousandth foot.   _____

13. Express 25.4 miles in feet.   _____

14. Express 1,156 inches in yards, feet, and inches.   _____

15. Express 2½ miles in yards.   _____

16. What part of a mile is 1,000 feet? Express the answer to the nearest hundredth.   _____

17. Express 66 inches in feet.   _____

18. Express 900 inches as yards.   _____

19. A workbench is 34 inches wide and 8 feet long. If a sheet of particle board 4 feet by 8 feet is purchased to cover it, what width is the piece that is left over?   _____

20. Speedometers register in tenths of a mile. How many feet is eight-tenths of a mile?   _____

Unit 23 METRIC MEASUREMENTS

BASIC PRINCIPLES OF METRIC MEASUREMENT

The metric system is used on all automobiles produced today. A mechanic must be able to measure in metric as well as the English system.

The metric system is not difficult to understand if you remember that it is based on *multiples of ten,* the same as *our monetary system.* The prefix of each unit tells what fraction or multiple of a meter is being used.

milli	=	1/1000 of a meter (or metre)	deci	=	1/10 of a meter
centi	=	1/100 of a meter	kilo	=	1000 meters

Another way of understanding the metric system is to remember the following spectrum of metric prefixes.

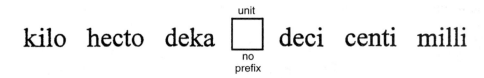

In the center of the spectrum is unit. Units in the metric system are grams, liters, and meters. Whenever you need to convert a metric measurement into another size of the same unit, you will move the decimal point to the right or left and add zeros until you arrive at the desired unit.

Example: Convert 2 meters into centimeters.

Step 1.   Recall that 2 meters is the same as 2.0 meters.

Step 2.   Place 2.0 in the unit box.

Step 3.   Count the number of metric prefixes you need to move to the right to get to the centimeter prefix. In this case, it is 2 prefixes away.

Step 4.   Move the decimal to the right 2 places: 2.0 to 2 0 0.0. Read the new number with its new prefix: 200.0 centimeters.

What if you were asked to convert 200 centimeters into meters? We already know the answer from the above problem; however, you would place 200.0 over the centimeter prefix and move the decimal point to the left two spaces. Therefore, 200 cm is equal to 2 meters.

Example: Convert 10 centimeters into millimeters.

Step 1.   Recall that 10 centimeters is the same as 10.0 centimeters.

Step 2.   Place 10.0 over the centimeter prefix.

Step 3.   Count the number of metric prefixes you need to move to the right to get to the millimeter prefix. In this case, it is 1 prefix away.

Step 4.   Move the decimal to the right 1 place: 10.0 to 1 0 0.0. Read the new number with its new prefix: 100.0 millimeters.

Example: Convert 28 millimeters into centimeters.

Step 1.   Recall that 28 millimeters is the same as 28.0 millimeters.

Step 2:   Place 28.0 over the millimeters prefix.

Step 3:   Count the number of metric prefixes you need to move to the left to get to the centimeters prefix. In this case, it is 1 prefix away.

Step 4:   Move the decimal to the left 1 place: 28.0 to 2.8. Read the new number with its new prefix: 2.8 centimeters.

Example: Convert 2,000 grams to kilograms. Place 2,000 above unit, and move its decimal point (2,000.0) to the left three places: 2,000.0 grams = 2.0 kilograms.

Example: Convert 3 liters into milliliters. Place 3 above unit, and move its decimal point three places to the right: 3,000 milliliters.

The following conversion table can be used to convert from the metric system to the English system, or to convert from English to metric.

TABLE OF LINEAR EQUIVALENTS

1 inch	=	0.0254 meter	1 meter	=	39.37 inches
1 inch	=	0.254 decimeter	1 decimeter	=	3.937 inches
1 inch	=	2.54 centimeters	1 centimeter	=	0.394 inch
1 inch	=	25.40 millimeters	1 millimeter	=	0.039 inch
		1 foot	=	0.304 8 meter	
		1 yard	=	0.914 4 meter	

Example: To change 2 meters to inches, divide by .0254 (the number of meters in 1 inch).

$$\frac{2 \text{ meters}}{.0254} = 78.74 \text{ inches}$$

Example: To change 100 inches to meters, divide by 39.37 (the number of inches in 1 meter).

$$\frac{100 \text{ inches}}{39.37} = 2.54 \text{ meters}$$

PRACTICAL PROBLEMS

1. How many centimeters are in 1 meter? _____

2. How many millimeters are in 1 meter? _____

3. How many millimeters are in 2½ meters? _____

4. How many centimeters are 275 millimeters? _____

5. How many millimeters are in 26 centimeters? _____

6. Express 6 centimeters to the nearest hundredth inch. _____

7. Express 37 millimeters to the nearest hundredth inch. _____

8. Find, in inches, the diameter of an 18-millimeter spark plug. _____

9. A generator pulley is 4 inches in diameter. What is the diameter in millimeters? _____

10. One size International Standard thread has a diameter of 22 millimeters and a pitch of 2½ millimeters.

    a. What is the diameter in inches? Express the answer to the nearest thousandth inch. a. _____

    b. How many threads per inch are there? Express the answer to the nearest whole thread. (Hint: Express pitch in inches, and use the formula at 1/Pitch = No. of threads.) b. _____

11. How many threads per centimeter does a ¼"–28 NF screw have? Express the answer to the nearest whole thread. _____

12. What is the diameter of a ¼"–28 NF screw in millimeters? _____

13. The cylinders of a foreign automobile are honed to fit an 80-millimeter piston allowing 0.002-inch clearance. What is the inside micrometer setting to the nearest thousandth inch? _____

14. A mechanic needs a 0.2-millimeter feeler gauge to adjust valves but only has an English feeler gauge in thousandths. What thickness in thousandths should the mechanic use? _____

15. To loosen a 19-millimeter lug bolt, a mechanic who does not have a metric wrench can use an English wrench that measures: _____

16. A brake drum on a late model car is marked 24.13 centimeters. What reading should the mechanic read on a brake drum micrometer calibrated in thousandths? _____

17. The diameter of a tire is 205 millimeters. What is the diameter in inches? _____

18. The minimum thickness of a brake rotor is 28.45 millimeters. How many thousandths would it measure with an English micrometer? _____

19. Toe-in on a car is ⅛ inch. What is the metric equivalent? _____

20. A radiator hose measuring 5 centimeters inside diameter has to be replaced. What size hose that is sold with English dimensions can be used? _____

21. A disc brake rotor marked with a minimum thickness of 1.13 centimeters is measured with an English micrometer and found to be 0.465 inch thick. Can it be used? _____

22. The crankshaft of a front-wheel drive, four-cylinder engine is listed as 16.83 inches long. What is the length in millimeters? _____

23. The tread of the rear wheels on a car is increased 80 millimeters from the year before. What is the increase in tread expressed in inches? _____

24. The overall length of a new car is listed as 14 feet 8.16 inches. Express the length in meters. _____

25.   In a certain model year, the sedan is listed as 4,655 millimeters long. The station wagon is 4,745 millimeters long. How many inches longer is the station wagon?   _____

26.   The braking distance from 60 mph to stop is 233 feet. How many meters is this?   _____

27.   A 10-millimeter wrench is a very popular size on metric automobiles. What is the English size wrench that is just a little smaller than a 10-millimeter wrench?   _____

28.   What is the English size wrench that is just a little larger than a 10-millimeter wrench?   _____

29.   Would an 18-millimeter socket fit a nut that measures ¾ inch across the flats?   _____

30.   A 15-millimeter wrench is just a little larger than what popular English size wrench?   _____

31.   A technician cannot get wheel cylinder cups for a Citroen that measure 2.5 centimeters. Is there an English size the technician can use?   _____

32.   A carburetor float level specification is given as 15 millimeters plus or minus 1 millimeter. The technician only has an English scale. What is the highest measurement the technician should use?   _____

33.   The head of a valve measures 35 millimeters. What is this in English measurement?   _____

 # Unit 24 SCALE READING

BASIC PRINCIPLES OF SCALE READING

The automotive technician must be able to make accurate measurements with both English and metric scales.

Study the diagram of an English scale before attempting the PRACTICAL PROBLEMS using English measurement.

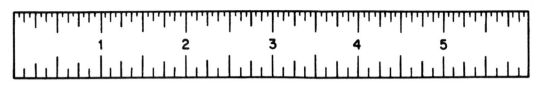

The scale used here has the upper part divided by 16 equal divisions to the inch ($\frac{1}{16}$"). The lower side is divided in eight equal divisions to the inch ($\frac{1}{8}$"). Reduce the fraction to its lowest terms, if necessary.

Example: A part that measures $\frac{6}{8}$" is reduced to $\frac{3}{4}$", and $\frac{18}{32}$" is reduced to $\frac{9}{16}$".

Some scales used in the automotive trade are graduated in divisions as small as 64 divisions per inch ($\frac{1}{64}$").

The increased use of metric measurements on American automobiles makes it important for a mechanic to be familiar with the metric scale.

On this metric scale, the numbered divisions are centimeters (cm). The unnumbered divisions are millimeters (mm). Ten millimeters equal one centimeter.

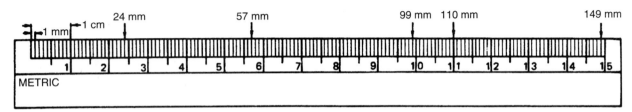

PRACTICAL PROBLEMS

Note: Use this ruler for problems 1 through 11.

1. On the upper part of the scale, there are 16 divisions to the inch. Each division equals $\frac{1}{16}$ inch. How many $\frac{1}{16}$ inches are there in 4 inches?   _____

2. There are $\frac{8}{8}$ in an inch. How many $\frac{1}{8}$ inches are there in the diameter of a 4-inch piston?   _____

3. How many $\frac{1}{16}$ inches are there in a $\frac{5}{8}$-inch diameter bolt?   _____

4. How many $\frac{1}{16}$ inches are there in a $\frac{3}{4}$-inch piston pin?   _____

5. How many $\frac{1}{16}$ inches are there in $\frac{1}{4}$ inch?   _____

6. How many $\frac{1}{16}$ inches are there in $\frac{7}{8}$ inch?   _____

7. How many $\frac{1}{8}$ inches are there in $\frac{3}{4}$ inch?   _____

8. What is the size of a socket $\frac{1}{8}$ inch larger than $\frac{1}{4}$ inch?   _____

9. What is the size of a socket $\frac{1}{16}$ inch larger than $\frac{1}{2}$ inch?   _____

10. What is the size of a nut $\frac{1}{4}$ inch smaller than $\frac{5}{8}$ inch?   _____

11. What is the diameter of a hose $\frac{1}{16}$ inch larger than $2\frac{3}{8}$ inches?   _____

Give, in inches, the scale readings indicated by numbers 12 through 20. Record each dimension in the space provided.

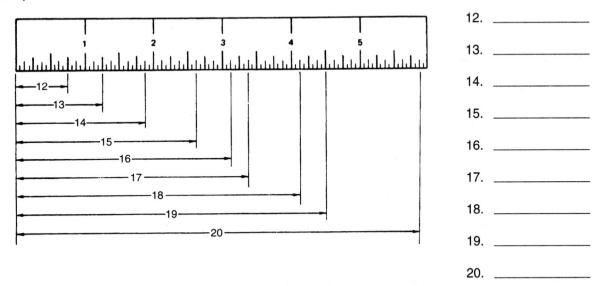

12. _____

13. _____

14. _____

15. _____

16. _____

17. _____

18. _____

19. _____

20. _____

Give, in millimeters, the scale readings indicated by numbers 21 through 25. Record each dimension in the space provided.

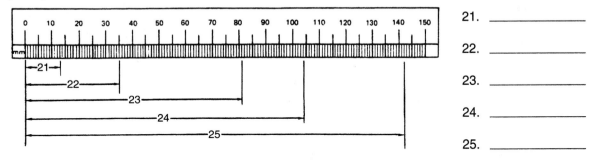

21. _____

22. _____

23. _____

24. _____

25. _____

Give, in centimeters, the scale readings indicated by numbers 26 through 30. Record each dimension in the space provided.

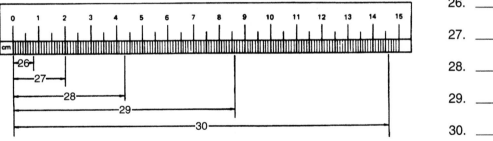

26. _____

27. _____

28. _____

29. _____

30. _____

Using a scale, measure and give the length in inches. (Note: Measure to the nearest $\frac{1}{16}$ inch.)

31. |—————————| _____

32. |————————————| _____

33. |———————————————| _____

34. |————| _____

35. |————————————| _____

36. |———————————————| _____

Using a metric scale, measure and give the length in millimeters.

37. |————————————| _____

38. |—————————| _____

39. |————————————————————————| _____

40. |——————————| _____

41. |——| _____

42. |—————————————| _____

Note: Use this diagram for problems 43 through 46.

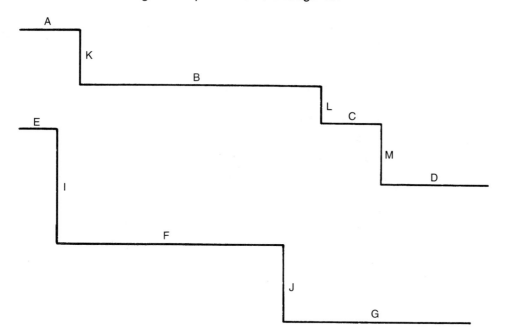

43.   Measure and give the total length of A + B + C + D in inches.   _____

44.   Measure and give the total length of E + F + G in centimeters.   _____

45.   Measure and give the total length of I + J in millimeters.   _____

46.   Measure and give the total length of K + L + M in inches.   _____

Unit 25 SCALE READING OF TEST METERS

BASIC PRINCIPLES OF READING TEST METERS

In this era of computer technology, many automotive meters are not digital. Technicians are required to use both digital and analog (a scale with a movable needle) type meters. New uses for older analog meters have been introduced by manufacturers. An example is found in problem 4, in which an older dwell meter on the six-cylinder scale is used to see if the computer is controlling the system.

When reading a meter, it is very important that you be directly in front of it. If you are reading it from an angle, you will get an incorrect reading. This problem is called *parallax.*

PRACTICAL PROBLEMS

1. A technician checks point adjustment with the dwell meter pictured here. What is the number of degrees of dwell on this engine?   _____

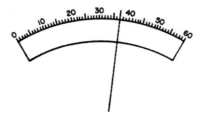

2. Using a voltmeter to check the battery pictured here, what is the reading to the nearest tenth of a volt?   _____

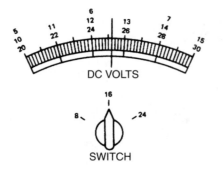

3. To test the secondary circuit of an ignition coil, an ohmmeter is connected as shown here. What is the number of ohms resistance in the coil? Note the position of the switch.   _____

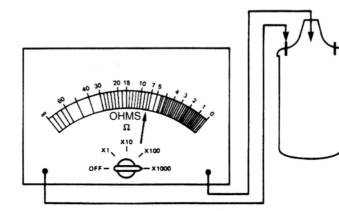

4. On a car equipped with a computer, a dwell meter on the six-cylinder scale is used to see if the mixture is too rich or too lean. The needle on the meter swings back and forth. This is called *varying*. What are the readings the needle is varying on the dwell meter shown?

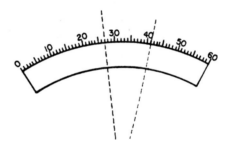

    _____° to ____°

5. To test the primary circuit of an ignition coil, an ohmmeter is connected as shown here. What is the number of ohms resistance in the coil? Note the position of the switch.   _____

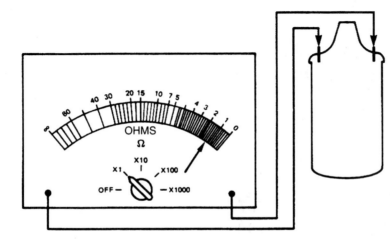

6. Testing a computer car as in problem 4, the needle is varying, showing that the computer is in closed loop. What are the readings the needle is varying?

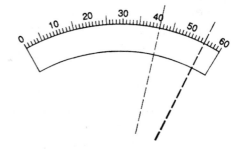

_____° to _____°

7. A technician is testing the emissions from the exhaust of a car. The infrared exhaust analyzer has two meters with two scales each. The black button tells which scale to read. If the wrong scale is read, the diagnosis will be incorrect. Read the two meters shown.

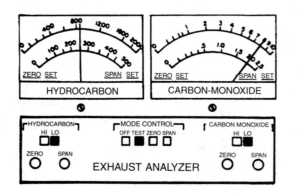

_____
ppm Hydrocarbons

_____
% Carbon Monoxide

8. Sometimes an automotive technician is called upon to read two meters at the same time. Such is the case when load testing a battery. To make it more difficult, the one meter reading is coming down while the other is going up. As the load is decreased, the amperes go up and the voltage drops. What are the two readings on the meter?

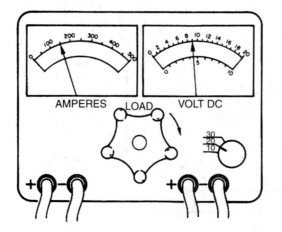

_____ Volts

at _____ Amps

9. What are the two readings on the meters in this battery load test?

_____ Volts

at _____ Amps

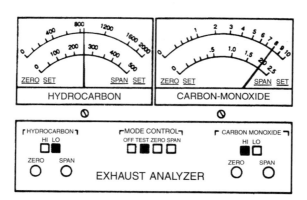

10. The technician is testing the exhaust emissions of a car that is running very smoothly but is getting poor gas mileage. The black button tells which scale to read.

_____

ppm Hydrocarbons

_____

% Carbon Monoxide

11. The technician is testing a car that is missing badly. Read the two meters.

_____

ppm Hydrocarbons

_____

% Carbon Monoxide

12. A technician is testing the total advance of a distributor with a meter on the timing light at 2,500 revolutions per minute (rpm). What is the number of degrees of advance?

_____

13. After making the total advance test in problem 12, the technician removes the vacuum line to the distributor so that only the mechanical advance will be tested. What is the reading of the mechanical advance test?

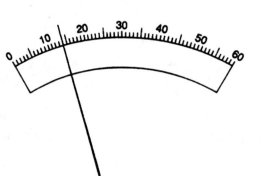

_____

In the following questions, read the voltmeter to the nearest tenth of a volt. Be sure to note the position of the switch in each case.

14.

15.

16.

17.

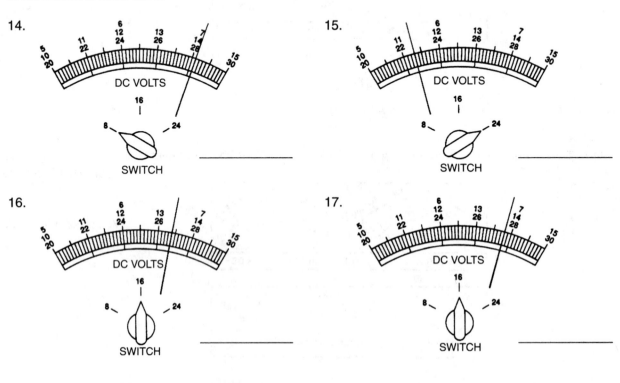

Unit 26 CIRCULAR MEASUREMENT

BASIC PRINCIPLES OF CIRCULAR MEASUREMENT

A mechanic must frequently make circular measurements.

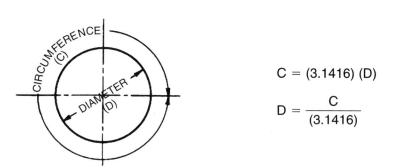

C = (3.1416) (D)

$$D = \frac{C}{(3.1416)}$$

Note: The circumference is the distance around a circle. The diameter is the distance across a circle. Sometimes in the automotive trade the diameter is referred to as the *bore.* (The radius is half of the diameter.) The length of the circumference of any circle is 3.1416 times the diameter. (Answers will vary if $3\frac{1}{7}$ is used instead of 3.1416.) The mathematical symbol for this value (3.1416) is called "*pi.*" *Pi* is represented by the Greek letter π.

PRACTICAL PROBLEMS

Note: If C = (3.1416) (D) is used, round the answer to the nearest thousandth.

1. If the diameter of a circle is 3 inches, find the circumference. _____

2. Find the circumference of a $4\frac{7}{16}$-inch diameter circle. _____

   Note: Use this diagram for problems 3 and 4.

3. What is the circumference of this fan pulley? _____

4. What is the circumference of this crankshaft pulley? _____

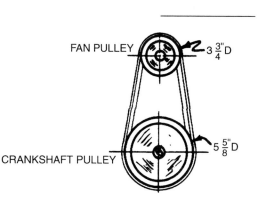

FAN PULLEY — $3\frac{3}{4}"$ D

CRANKSHAFT PULLEY — $5\frac{5}{8}"$ D

5. In one turn of a speed wrench, the hand makes a circle. What is the diameter of the circle made by this speed wrench?   _____

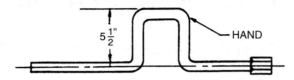

$5\frac{1}{2}$"   —— HAND

6. The circumference of a circle is $2\frac{1}{6}$ inches. What is the diameter?   _____

7. What is the diameter of a circle with a circumference of $5\frac{1}{2}$ inches?   _____

8. A 6-inch wire is bent into the shape of a circle. What is the diameter of the circle?   _____

9. The pitch circle of the gear is $5\frac{1}{4}$ inches in diameter, and it has 24 teeth. What is the distance between teeth measured on the circumference of the pitch circle?   _____

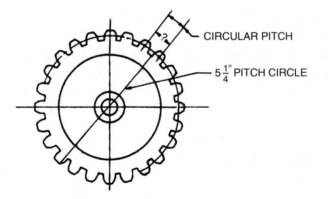

CIRCULAR PITCH

$5\frac{1}{4}$" PITCH CIRCLE

10. The face of this 5-inch diameter gauge is divided in 8 equal parts. What is the length of each division measured on the circumference of the gauge?   _____

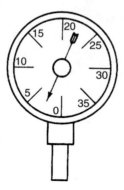

Unit 27 ANGULAR MEASUREMENT

BASIC PRINCIPLES OF ANGULAR MEASUREMENT

Degrees, minutes, and seconds are used for angular measurements in both the English and metric systems.

Caution: Do not confuse the symbols for minutes and seconds with the symbols for feet and inches, as they look the same.

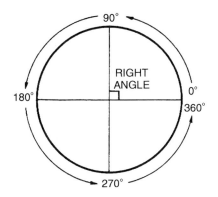

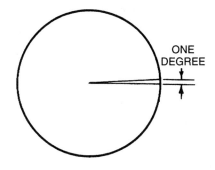

1 circle = 360°
¼ circle = 90°
½ circle = 180° = semicircle

1 degree (1°) = 60 minutes (60')
1 minute (1') = 60 seconds (60")

PRACTICAL PROBLEMS

1. How many degrees are in a semicircle?   _____

2. How many degrees are in a quarter of a circle?   _____

3. What part of a circle is a 60° angle?   _____

4. What part of a circle is a 45° angle?   _____

5. What part of a circle is a 30° angle?   _____

6. A *regular hexagon* has six equal sides and six equal angles. A regular hexagon is drawn inside (inscribed in) this circle. The radii form equal angles at the center. Find the number of degrees in each angle at the center.

   _____

7. A *regular octagon* has eight sides and eight equal angles. Inscribed in this circle is a regular octagon. The radii form equal angles at the center. How many degrees are in each angle at the center?

   _____

8. A *straight angle* contains 180 degrees. How many degrees are in angle **A**?

   _____

9. In this square, angle **A** is a right angle. The diagonal of the square (dimension **X**) divides angle **A** into two equal angles. Find the number of degrees in each of those angles.

   _____

10. The circumference of this circle is 6 inches. In making a gasket, a mechanic lays off 12 equally spaced holes on the circumference. Find the number of degrees between the centers of the holes.

    _____

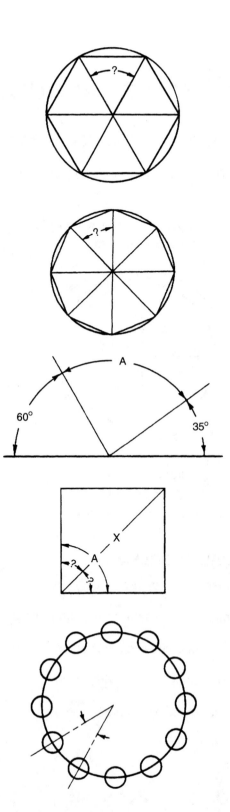

11. On the circumference of a 5½-inch diameter circle, 15 equally spaced holes are laid off. How many degrees are between the centers of the holes?   _____

12. Five equally spaced holes are drilled around a circle. Find the number of degrees between the centers of the holes.   _____

13. Seven holes drilled around a circle are equally spaced. How many degrees, minutes, and seconds are between the centers of the holes? Express the answer to the nearest second.   _____

14. How many degrees are between each of the distributor cam lobes in an 8-cylinder engine?   _____

15. How many degrees are between each of the distributor cam lobes in a 6-cylinder engine?   _____

16. Two revolutions of the crankshaft are needed to complete the cycle (intake, compression, power, and exhaust) in an engine. How many degrees will the crank turn between cylinder firings in an 8-cylinder engine?   _____

17. How many degrees will the crankshaft turn between cylinder firings in a 6-cylinder engine?   _____

18. The intake valve opens when the crank is 13 degrees before top dead center (TDC) and closes 48 degrees after bottom dead center (BDC). How many degrees does the crank move on the *intake stroke*?

_____

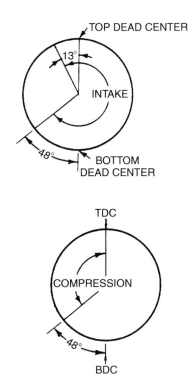

19. All valves are closed from 48 degrees after bottom dead center to exact top dead center. How many degrees does the crank move during the *compression stroke*?

_____

20. The crank travels through 134 degrees on the *power stroke.* How many degrees before bottom dead center does the exhaust valve open?

_____

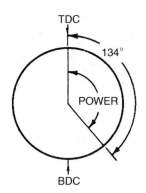

21. The exhaust valve opens at the end of the power stroke and remains open until the crank is 13 degrees after top dead center. How many degrees does the crank move during the *exhaust stroke*?

_____

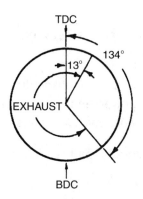

22. What percent of the entire cycle of strokes is the power stroke to the nearest tenth? Use a 134-degree power stroke. (Remember, it takes two revolutions to complete the entire cycle.)

_____

Note: In questions 23 through 26, use the formula $C = \pi D$. Round to the nearest thousandth.

23. A certain engine is timed when the inlet valve opens (8 degrees before top dead center), but there is no mark on the flywheel to show this position. What distance on the circumference of the 18-inch diameter flywheel is measured from the line of top dead center to get the proper position of the crank?

_____

24. How many inches are represented by 5 degrees on the rim of a 16-inch diameter flywheel?

_____

25. On the rim of a 21-inch diameter flywheel, how many inches are represented by 20 degrees?

_____

26. A flywheel is $14\frac{1}{2}$ inches in diameter. How many inches are represented by 40 degrees on the rim?

_____

Unit 28 AREA AND VOLUME MEASUREMENT

BASIC PRINCIPLES OF AREA AND VOLUME MEASUREMENT

Area is surface measurement that has length and width or length and height but no thickness. It is usually found by multiplying length by width or height. The answer is expressed in square units. Change both measurements to the same linear units before multiplying.

Example: Area = length × width
                A = 3" × 2"
                A = 6 square inches

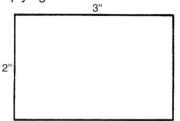

Volume is the space occupied by a body and is the product of three linear measurements. It is found by multiplying length by height by width. Change all measurements to the same linear units before multiplying. The answer is always expressed in cubic units.

Example: Volume = length × height × width
                V = 3" × 2" × 2"
                V = 12 cubic inches

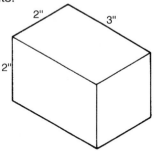

The volume of a cylinder is determined by finding the area of the circle and multiplying it by the height.

Example: To find the volume of a cylinder with a diameter of 8 centimeters and a height of 8 centimeters:
                $V = \pi \times r^2 \times H$
                $V = 3.14 \times 4 \times 4 \times 8$
                V = 402.12 cubic centimeters (cc)

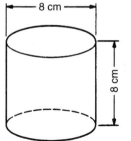

Reminder: R-radius is equal to ½ of D-diameter.

Today technicians are confronted daily with engine displacement expressed in cubic centimeters or liters. Remember that there are 1,000 cubic centimeters in a liter, so a 2.3-liter engine can also be called 2300 cubic centimeters. It is helpful to remember that there are *61 cubic inches in a liter.*

PRACTICAL PROBLEMS

1. How many square feet are there in a parking area 150 feet × 200 feet? _____

2. An acre contains 43,560 square feet. What percent of an acre is the area of a 150 feet × 200 feet parking lot? Express the answer to the nearest tenth percent. _____

3. Cork gasket material is bought by the square foot. The roll is 30 inches wide. If 25 square feet of material are bought, what is the length in feet? _____

4. A piece of cork 2½ feet long and 14 inches wide is needed for each truck side cover gasket. The roll of cork gasket material is 12 feet long and 30 inches wide. How many gaskets can be cut from this roll as shown? _____

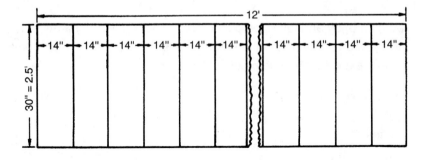

5. How many square inches are there in a 14 inches wide × 2½ feet long gasket? _____

6. Fiberglass costs $3.65 per square foot. What is the cost, to the nearest cent, of a piece 42 inches × 3 feet? _____

7. A mechanic relines a truck body with flat stock. The floor measures 8 feet × 10 feet, two sides measure 6 feet × 10 feet each, and the front measures 6 feet × 8 feet. What is the total area covered in square feet? _____

8. A storage garage has an area of 12,000 square feet. A space 8 feet × 12 feet is allowed for each car in the garage. How many cars can be parked in the garage? _____

9. A 6-cell storage battery has 8 positive plates per cell. Both sides of the plates are active. Each plate measures 5½ inches × 6¼ inches. How many square inches of active material are in this cell?

    _____

10. A 30 feet × 40 feet garage is rented monthly. The owner charges $0.55 per square foot. What is the rent per month?

    _____

11. A mechanic is building a garage. Spaces are needed for: a grease rack, 10 feet × 20 feet; motor repair for 3 cars, 20 feet × 30 feet; a bench, 2 feet × 30 feet; a wash rack, 10 feet × 20 feet; and storage, 10 feet × 13 feet. What is the cost at $39.50 per square foot?

    _____

    Note: Use this information for problems 12 through 23.

Specific gravity refers to the weight of a substance compared to an equal quantity of water. For example, sulfuric acid has a specific gravity of approximately 1.835. This means that acid is 1.835 times as heavy as an equal volume of water. The weight of a gallon of a substance is the specific gravity times the weight of one gallon of water. This can be written as:

$$W = \text{s.g.} \times \text{wt. of water}$$

For example, the weight of one gallon of acid is:

$$W = 1.835 \times 8.333 \text{ pounds} = 15.291 \text{ pounds (approx.)}$$

Note: Use this chart for problems 12 through 23.

Water	=	1.000 specific gravity
Acid	=	1.835 specific gravity
Alcohol	=	0.816 specific gravity
Gasoline	=	0.728 specific gravity
1 U.S. gallon	=	231 cubic inches
1 gallon	=	4 quarts
2 pints	=	1 quart
1 cubic foot of water	=	62.5 pounds
1 U.S. gallon of water	=	8.333 pounds
1 liter of water	=	1 kilogram
1 kilogram	=	1,000 gram
1 kilogram	=	2.2 pounds

12. What is the weight of a gallon of gasoline to the nearest hundredth?

    _____

13. Find the weight of 12 pints of water.   _____

14. What is the weight of the contents of a 50-gallon drum of alcohol? Express
     the answer to the nearest hundredth pound.   _____

15. Sulfuric acid for storage batteries is sometimes shipped in 200-pound
     carboys (200 pounds of liquid). If the specific gravity is 1.835, how many
     gallons will this carboy hold? Express the answer to the nearest tenth gallon.   _____

16. In a certain locality, the legal load limit of a 1-ton delivery truck is 2,000
     pounds. A 550-gallon tank weighing 185 pounds is mounted on a 1-ton truck.
     The tank is filled with gasoline. Find, to the nearest tenth pound, the amount
     by which the load exceeds the legal limit.   _____

17. Find, to the nearest hundredth, the number of gallons in a cubic foot of water.   _____

18. How many gallons does a cubic foot of gasoline contain? Express the answer
     to the nearest hundredth.   _____

19. A dump truck is carrying 2 cubic yards of loam. The weight of loam per cubic
     foot is 125 pounds. Find the weight of this load in pounds.   _____

20. Ford made an engine that had a displacement of 427 cubic inches. Express
     this in liters and cubic centimeters.   _____ liters

                                                                    _____ cc

21. What is the weight of 82 liters of gasoline? (Hint: First find the weight of a liter
     from the table, then multiply by the specific gravity of gasoline. Multiply by
     82.)   _____

22. What is the volume of a cylinder that has an area of 80 centimeters and a
     height of 30 centimeters? Give the answer in liters.   _____

23. How would a 1½-ton truck be rated in kilograms?   _____

Unit 29 TIME, SPEED, AND MONEY CALCULATIONS

BASIC PRINCIPLES OF TIME, SPEED, AND MONEY CALCULATIONS

An automotive technician is often required to solve problems involving time and money as they relate to the technician's pay.

Many mechanics are paid by the hour, but others work under the flat-rate system. The flat-rate system means that each job has been time-studied, and a time has been allowed for that job. If a mechanic finishes in less time, the mechanic is still paid the flat rate. (If the job takes longer, the mechanic still gets the same pay for the job.) Flat-rate times are given in hours and tenths of an hour; one-tenth is six minutes.

PRACTICAL PROBLEMS

1. A mechanic works on a differential overhaul. The starting time is 8:27 A.M., and finishing time is 3:10 P.M. At $12.85 per hour, find, to the nearest cent, the amount the mechanic earned. (Allow one hour for lunch.) _____

2. The flat-rate price for the labor on a job is $189.00. The job takes 6 hours. The mechanic receives 50% of the flat rate. What are the average hourly earnings for this job? _____

3. An assembly job takes 20 minutes. How many jobs is a worker expected to do in an 8-hour day? _____

4. A job takes 102 hours to finish. Using an 8-hour workday, in how many days and hours is the job finished? _____

5. Find the number of weeks, days, and hours in 210 hours. (Use an 8-hour day and a 40-hour week.) _____

6. Express 0.6 hour in minutes. _____

7. A worker is 23 minutes late. Find, to the nearest tenth hour, this time. _____

8. It takes 30 minutes to drive 10 miles. What is the average rate of speed in miles per hour? _____

9. If 45 miles are traveled in 50 minutes, what is the average rate of speed in miles per hour?   _____

10. In 1 hour and 45 minutes, 120 kilometers are driven. What is the average rate of speed in kilometers per hour? Express the answer to the nearest tenth.   _____

11. A 288-kilometer trip takes 4.6 hours. Find, to the nearest tenth, the average rate of speed in kilometers per hour.   _____

12. An airplane travels at a speed of 350 miles per hour. What is the plane's mileage per minute to the nearest hundredth?   _____

13. Express 60 miles per hour to the nearest hundredth feet per second.   _____

    Note: Use this information for problems 14 through 17.

    Knot means nautical miles per hour. One *nautical mile* = 1.1515 land miles. One *land mile* = 0.8643 nautical miles.

14. A boat has a speed of 28 knots. Express this speed in miles per hour.   _____

15. What is the speed in kilometers per hour of a ship with a top speed of 32 knots? (One kilometer = 0.625 land mile.) Round the answer to the nearest hundredth.   _____

16. Express a land speed of 50 miles per hour to the nearest tenth knot.   _____

17. A boat travels at a land speed of 60 miles per hour. How many knots is it traveling? Express the answer to the nearest tenth.   _____

    Note: Use this diagram for problems 18 through 20.

The engine makes 2 revolutions to every 1 revolution of the distributor.

18. A V-8 engine is turning over at 3,600 revolutions per minute (rpm). How many times will the plugs fire in 1 minute?   _____

19. An engine is running 3,400 rpm and has a 4-stroke. What is the average speed of a piston in feet per minute? Express the answer to the nearest hundredth.   _____

20. The primary circuit in the distributor of a 4-cylinder engine is interrupted 15 times per second. What is the number of revolutions per minute for the flywheel?   _____

Ratio and Proportion

SECTION

6

Unit 30 RATIOS

BASIC PRINCIPLES OF RATIOS

A ratio is a comparison of one quantity with another quantity that is similar. The automotive technician frequently uses ratios to describe how two units compare with another.

Examples: Diameters of pulleys
Rpm of one unit to another
Number of gear teeth
Turns of wire
Volume of a cylinder

After stating the like quantities, it is necessary to reduce the ratio to lowest terms.

Example: Two windings of a coil with 1,600 turns to 400 turns would be expressed as 1,600:400, which would be reduced to 4:1.

PRACTICAL PROBLEMS

Gear ratio compares the number of teeth on different gears. It is the comparison of the number of teeth on the *driven gear* to the number of teeth on the *driver gear.* To find ratio, always divide the larger number by the smaller number. The answer, called the quotient, is then compared to one. If the gear ratio of **A** to **B** is 2:1, gear **A** has twice as many teeth as gear **B**.

   Note: Use this diagram for problems 1 through 3.

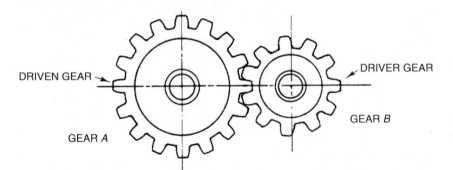

DRIVEN GEAR

DRIVER GEAR

GEAR *B*

GEAR *A*

1. Gear **A** has 60 teeth, and gear **B** has 20 teeth. What is the ratio of teeth on **A** to teeth on **B**?   _____

2. The driven gear has 75 teeth. The driver gear has 10 teeth. Find the ratio of the teeth on the driven gear to the driver gear.   _____

3. Gear **A** has 45 teeth, and gear **B** has 25 teeth. What is the ratio of the number of teeth on **A** to the number of teeth on **B**?   _____

On pulleys, the ratio of the diameters can be found. The ratio is the diameter of the *driven pulley* compared to the diameter of the *driver pulley*. If the ratio of **A** to **B** is 3:1, the diameter of pulley **A** is three times the diameter of pulley **B**.

   Note: Use this diagram for problems 4 through 6.

4. Pulley **A** is 35 inches in diameter, and pulley **B** is 7 inches in diameter. What is the ratio of the diameter of **A** to the diameter of **B**?   _____

5. The driven pulley is 15 inches in diameter. The driver pulley is 8 inches in diameter. What is the ratio of the diameters of the driven pulley to the driver pulley?   _____

6. The circumference of pulley **A** is 15.7078 inches. The circumference of pulley **B** is 7.8539 inches. Find the ratio of the circumference of **A** to the circumference of **B**.   _____

   Note: A gear ratio of the teeth on **A** to **B** is 2:1. This means that gear **B** moves twice as fast as gear **A**. A gear ratio of 3:1 means that gear **B** moves three times as fast as gear **A**.

7. The gear ratio of the teeth on **A** to **B** is 7:1. How many times faster does gear **B** move than gear **A**?   _____

Note: Use this information for problems 8 and 9.

On pulleys, the ratio of the diameters of **A** to **B** is 2:1. This means that pulley **B** turns two times for one turn of pulley **A**.

8. The ratio of the diameter of pulley **A** to **B** is 3:2. How many times does pulley **B** turn for one turn of pulley **A**?   _____

9. Pulley **B** turns four times for every turn of pulley **A**. What is the ratio of the diameter of **A** to the diameter of **B**?   _____

Note: Use this diagram for problems 10 and 11.

Power flowing through a train of gears flows
from the *driver gear* to the *driven gear*.

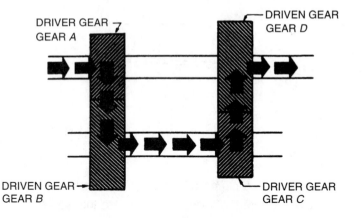

10. The driver gear, gear **A**, has 25 teeth. The driven gear, gear **B**, has 35 teeth. Find the gear ratio of the teeth on **B** to the teeth on **A**.   _____

11. Gear **C** has 15 teeth and gear **D** has 30 teeth. Find the ratio of the teeth on the driven gear to the teeth on the driver gear.   _____

Compression ratio is a comparison between the amount of space (cubic units) in the cylinder when the piston is at the bottom of the stroke, and the amount of space when the piston is at the top of the stroke. If there is eight times as much space when the piston is at the bottom of the stroke as when the piston is at the top of the stroke, the compression ratio is 8:1.

Note: Use this diagram for problems 12 through 14.

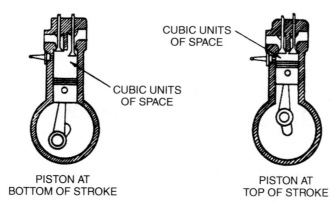

CUBIC UNITS OF SPACE

CUBIC UNITS OF SPACE

PISTON AT
BOTTOM OF STROKE

PISTON AT
TOP OF STROKE

COMPRESSION RATIO = CUBIC UNITS (BDC) : CUBIC UNITS (TDC)

12. There are 33 cubic inches of space when the piston is at the bottom of the stroke. When the piston is at the top of the stroke, there are 4 cubic inches of space. What is the compression ratio? _____

13. When the piston is at the bottom of the stroke, there are 43 cubic inches of space. There are 5 cubic inches when the piston is at the top of the stroke. What is the compression ratio? _____

14. There are 35.01 cubic inches of space when the piston is at the bottom of the stroke. When the piston is at the top of the stroke, there are only 4.5 cubic inches of space. Find the compression ratio. _____

Transmission gear ratio refers to the number of times the speed is reduced by the transmission. It compares the speed of the crankshaft to the speed of the driveshaft. A transmission ratio of 3:1 means the speed is reduced three times by the transmission, that is, the rpm of the crankshaft is three times the rpm of the driveshaft.

In finding a transmission gear ratio, first find the gear ratio of each set of gears in mesh: the number of teeth on the *driven* gear compared to the number of teeth on the *driver* gear. Remember that power flows from the *driver* gear to the *driven* gear. The transmission gear ratio will be the product of the first terms of the gear ratios compared to the product of the second terms of the gear ratios.

Note: Use these diagrams for problems 15 through 18.

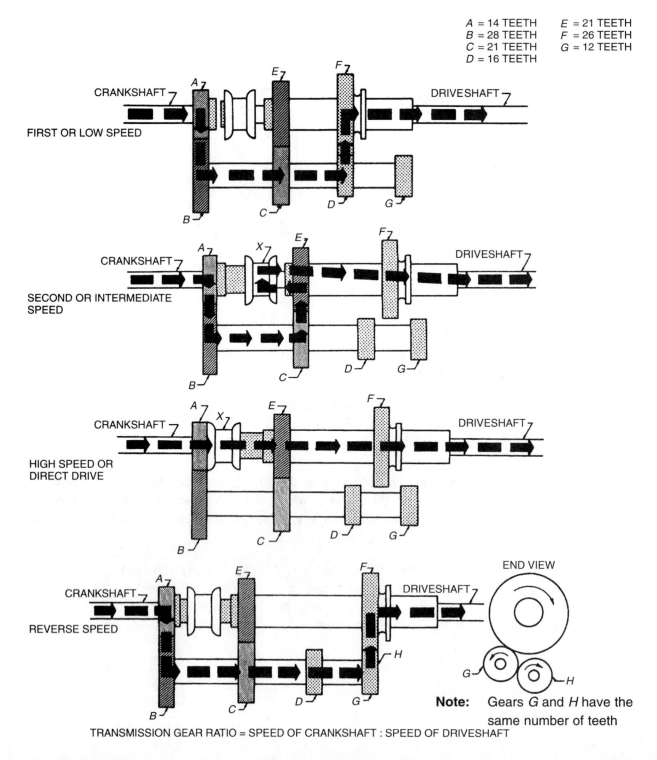

TRANSMISSION GEAR RATIO = SPEED OF CRANKSHAFT : SPEED OF DRIVESHAFT

15. In low speed (first), gears **A** and **B**, and gears **D** and **F**, are in mesh. What is the transmission gear ratio in low speed? _____

16. In intermediate speed (second), gears **A** and **B**, and gears **C** and **E**, are in mesh. What is the transmission gear ratio in intermediate speed? _____

17. In high speed, the internal splines in sleeve **X** mesh with teeth on the back of gear **A**. This gives a direct drive. What is the transmission gear ratio in high speed? _____

18. In reverse, gears **A** and **B**, and gears **G** and **F**, are in mesh. What is the transmission gear ratio? (The reverse idler changes direction of rotation but does not affect the speed.) Express the answer as a fraction. _____

Axle or *differential ratio* refers to the number of times the speed is reduced by the ring gear and pinion. It compares the speed of the driveshaft to the speed of the rear axle shaft. A gear ratio of 4:1 means that the rpm of the driveshaft is four times as great as the rpm of the rear axle shaft. To find the axle ratio, find the gear ratio of the *ring gear* compared to the *pinion gear.*

   Note: Use this diagram for problems 19 through 21.

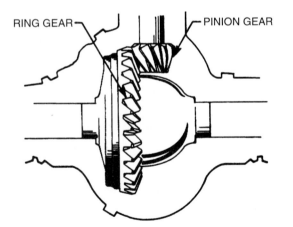

RING GEAR─

─PINION GEAR

AXLE OR DIFFERENTIAL RATIO = SPEED OF DRIVESHAFT: SPEED OF REAR AXLE SHAFT

19. The pinion gear has 12 teeth and the ring gear has 42 teeth. What is the axle ratio? _____

20. If the pinion gear has 14 teeth and the ring gear has 49 teeth, find the differential ratio. _____

21. A ring gear has 40 teeth, and a pinion gear has 16 teeth. What is the axle ratio? _____

Note: Use this information for problems 22 through 27.

Total gear reduction is a ratio of the rpm of the crankshaft to the rpm of the rear axle shaft. In measuring total gear reduction, both transmission gear ratios and differential (axle) ratios are used. For example, a transmission gear ratio is 2:1, and a differential ratio is 3:1. The total gear reduction ratio is 3 × 2:1 × 1, or 6:1. This means that the crankshaft turns six times while the rear axle shaft turns once.

22.   In low speed, the transmission gear ratio of a truck is 3.5:1, and the axle ratio
      is 4.7:1. What is the total gear reduction?   _____

23.   The second gear transmission gear ratio for a truck is 1.8:1. The rear axle
      ratio is 4.7:1. Find the total gear reduction.   _____

24.   A car is in high-speed position. The pinion gear has 9 teeth, and the ring gear
      has 28 teeth. What is the total gear reduction?   _____

25.   In second speed, the transmission gear ratio is 2.2:1, and the axle ratio is
      3.10:1. What is the total gear reduction?   _____

26.   In low speed, the transmission gear ratio is 2.85:1, and the axle ratio is
      2.95:1. What is the total gear reduction?   _____

27.   On a car with a total gear reduction of 8:1, if the engine is turning at 3,500
      rpm, how fast is the axle turning?   _____

      Note: Use this information for problems 28 through 31.

To increase gasoline mileage, many cars now have a fourth or fifth gear that is an overdrive. An overdrive means that the driven gear has fewer teeth and will turn faster than the driving gear. Overdrive ratios are expressed as a number less than one to one.

28.   A car is in fifth speed (overdrive). The driver gear has 17 teeth, and the driven
      gear has 13 teeth. What is the overdrive ratio?   _____

29.   If the engine is turning at 2,800 rpm, what is the driveshaft speed when the
      car is in overdrive? The overdrive ratio is 0.7:1.   _____

30.   On a car with an overdrive ratio of 0.65:1, if the car is driven 17,000 miles in
      overdrive, what is the actual mileage on the spark plugs?   _____

31.   If the engine oil should be changed every 7,500 miles on a car with a three-
      speed or automatic transmission, how many more miles between oil changes
      can the owner drive the same model car equipped with an overdrive with a
      0.7:1 ratio? (Assume that all of the miles are driven in overdrive.)   _____

Note: Refer back to the compression ratio discussion before problem 12 for help with questions 32 through 34.

32. A 2.2-liter diesel engine in a late-model car has 577.35 cubic centimeters (cc) of volume at the bottom of the stroke. When the piston is at the top of the stroke, there are 25.10 cc of volume. What is the compression ratio? _____

33. A sport truck with a 2.3-liter turbocharged diesel engine has a cylinder volume at bottom dead center of 605 cc. At top dead center, the volume is 30 cc. What is the compression ratio? _____

34. A V-8 diesel has a cylinder volume of 712.5 cc at bottom dead center and 35 cc at top dead center. What is the compression ratio? _____

35. What is the ratio of domestic manufactured cars to imports sold by a dealer if in one month he sells 45 domestic cars and 16 imports? _____

36. In one year a dealer sells 1,268 cars. After the cars are out of the warranty period, 390 owners continue to have their cars serviced at the dealership. What is the ratio of those who do not continue with the dealer to those who do? _____

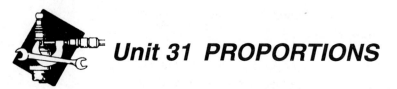 *Unit 31* **PROPORTIONS**

BASIC PRINCIPLES OF PROPORTIONS

A proportion is two equal ratios, one on each side of an equals sign.

Example: 3:1 = 12:4

The automotive technician can use the following formula to solve for an unknown value. Multiply the two outer numbers, and they will be equal to the product of the two inner numbers.

$$3:1 = 12:4$$

$$3 \times 4 = 12 \quad and \quad 1 \times 12 = 12$$

If three values are known, it is possible to solve for the unknown value by using elementary algebra.

Example: If a gear turns three revolutions while the second gear only turns one revolution, and the large gear has 12 teeth, we can find the number of teeth on the small gear as follows:

$$3:1 = 12:x$$

$$3x = 12$$

$$x = \frac{12}{3}$$

$$x = 4$$

PRACTICAL PROBLEMS

1. A gear with 40 teeth turning 200 rpm is in mesh with a gear with 10 teeth. Find the rpm of the small gear.   _____

2. A gear of 15 teeth turning 150 rpm is driving a gear of 25 teeth. Find the rpm of the driven gear.   _____

3. A 28-tooth gear running 320 rpm drives another gear. The other gear is running at 128 rpm. How many teeth does the driven gear have?   _____

4. Two gears have a gear ratio of 3.6:1. If the larger gear has 72 teeth, how many teeth does the smaller gear have?   _____

5. Two chain-driven sprockets have teeth as follows: smaller sprocket, 15 teeth; larger sprocket, 21 teeth. If the smaller sprocket runs at 1,939 rpm, find the rpm of the larger sprocket.

_____

6. A transmission gear ratio is 3.8:1. The crankshaft speed is 1,520 rpm. What is the rpm of the driveshaft?

_____

7. An axle ratio is 3.6:1. The driveshaft is turning 900 rpm. What is the rpm of the rear axle?

_____

8. The axle ratio is 4.5:1, and the pinion has 8 teeth. How many teeth does the ring gear have?

_____

9. A total gear reduction is 10.4:1. What is the rpm of the crankshaft if the rear wheels are turning 150 rpm?

_____

Note: Use this diagram for problems 10 through 14.

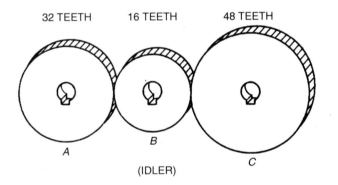

10. Gear **A** turns 300 rpm. Find the speed of gear **C**.

_____

11. If gear **C** meshes with gear **A** without the idler, what is the speed of gear **C**? (The speed of gear **A** is 300 rpm.)

_____

12. Does the idler cause any change in speed?

_____

13. If gear **A** rotates clockwise, in what direction does gear **C** rotate?

_____

14. Gear **C** meshes with gear **A** without the idler. Gear **A** rotates clockwise. In what direction does gear **C** rotate?

_____

Note: Use this diagram for problems 15 and 16.

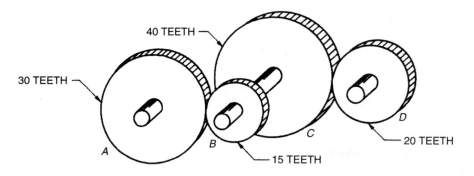

15. Gear **A** is the driver gear. The rpm of gear **A** is 300. What is the rpm of gear **D**?   _____

16. In the diagram, assume gear **D** is the driver gear and is turning 600 rpm. Find the rpm of gear **A**.   _____

Note: Use this diagram for problems 17 and 18.

A = 14 TEETH
B = 28 TEETH
C = 18 TEETH
D = 24 TEETH
E = 27 TEETH
F = 25 TEETH

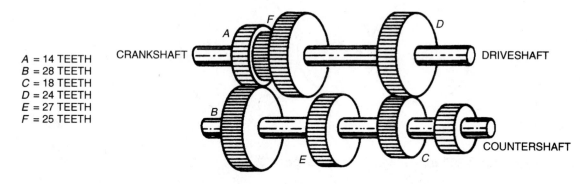

17. In this transmission, the crankshaft is turning 1,800 rpm. What is the rpm of the driveshaft? (Hint: The flow of power through the four gears is: **A B C D**.)   _____

18. In second speed in this transmission, gear **E** meshes with gear **F**. Gear **A** and gear **B** are still in mesh. What is the rpm of the driveshaft if the crankshaft is turning 1,800 rpm? (Hint: The flow of power through the four gears is: **A B E F**.)   _____

19. If the speed of pulley **A** is 500 rpm, find the speed of pulley **B**. _____

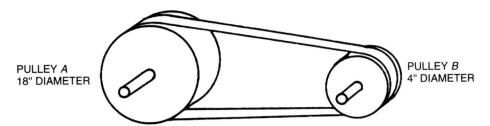

PULLEY *A*
18" DIAMETER

PULLEY *B*
4" DIAMETER

Note: Use this diagram for problems 20 and 21.

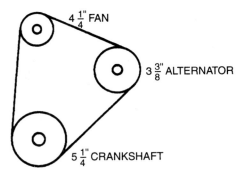

$4\frac{1}{4}$" FAN

$3\frac{3}{8}$" ALTERNATOR

$5\frac{1}{4}$" CRANKSHAFT

20. Find the rpm of the alternator pulley if the crankshaft turns 1,200 rpm. _____

21. Find the rpm of the fan pulley if the crankshaft turns 1,200 rpm. _____

22. The diameter of a driven pulley is 3 inches. The pulley is rotating at 350 rpm. The driver pulley is rotating at 1,500 rpm. Find the diameter of the driver pulley. _____

Note: Use this diagram for problems 23 and 24.

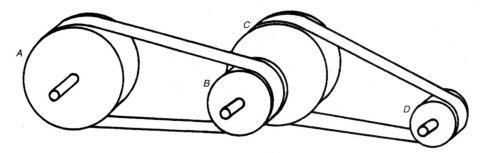

23.   Pulley **A** is 10 inches, **B** is 2¼ inches, **C** is 8¾ inches, and **D** is 3½ inches. The rpm of pulley **A** is 800. Find the rpm of pulley **D**.   _____

24.   Pulley **A** is 12 inches, **B** is 5 inches, and **C** is 15 inches. The rpm of pulley **A** is 1,000, and the rpm of pulley **D** is 2,500. Find the diameter of pulley **D**.   _____

Corresponding sides of similar triangles (triangles that have equal angles) are in proportion. In the following diagrams, for example:

$$\mathbf{a : A = b : B}$$
$$\mathbf{a : A = c : C}$$
$$\mathbf{b : B = c : C}$$

Note: Use these diagrams for problems 25 through 27.

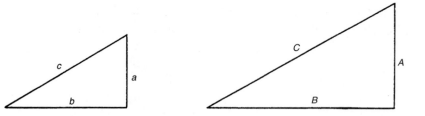

25.   If side **a** is 4 inches, side **b** is 7 inches, and side **B** is 21 inches, find side **A**.   _____

26.   In the larger triangle, side **A** is 10 feet and side **C** is 25 feet. In the smaller triangle, side **a** is 6 feet. Find, in feet, side **c**.   _____

27.   Find, in centimeters, side **b** when **B** is 8 centimeters, **C** is 20 centimeters, and **c** is 8 centimeters.   _____

28. The state automobile inspection permits headlights set at an angle. The drop in the light beam cannot be greater than 2 inches for each 25 feet measured horizontally. The headlight is 28 inches above the ground. How far ahead of the car is the roadway lighted? (Hint: Use similar triangles. The ratio of drop is 2 inches per 25 feet.)

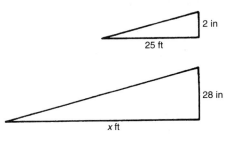

29. Car headlights are 30 inches above the road. They are tilted so that the drop in light beam is 4 inches per 25 feet. How far ahead of the car does the light beam hit the road? (Hint: Use similar triangles.)

30. Headlights that are 2 feet above the road light the roadway 500 feet ahead of the car. What rate of drop does the light beam have? Express the answer in inches per 25 feet.

31. The rate of drop in light beam is 1 inch per 25 feet. How high should the headlights be mounted to light the roadway 250 yards ahead of the car?

Note: Use this diagram for problems 32 through 35.

Taper is the difference in diameter between the two ends. *Standard taper pins* all have a taper equal to $\frac{1}{4}$ inch per foot.

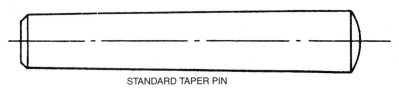

STANDARD TAPER PIN

32. What is the difference in diameters between the two ends if a taper pin is 4 inches long?

33. The large end of a standard taper pin *#000* measures 0.125 inch. The pin is 1 inch long. What is the measure of the small end of the pin? Express the answer as a fraction.

34. The small end of a *#3* standard taper pin measures 0.177 inch. What does the large end measure if the pin is 2 inches long? Express the answer to the nearest thousandth.   _____

35. What is the length of a standard taper pin if the large end measures 0.375 inch and the small end measures 0.3125 inch?   _____

36. A car travels 15 miles per hour (mph) when the crankshaft speed is 440 rpm. How fast is the car going when the crankshaft speed is 1,408 rpm?   _____

37. The speed of a car is 20 mph when the crankshaft speed is 670 rpm. How fast is the crankshaft turning when the speed of the car is 50 mph?   _____

38. The speed of the crankshaft in a car is 2,500 rpm when the car is traveling 75 mph. How fast is the car going when the crankshaft speed is 1,250 rpm?   _____

    Note: Use this information for problems 39 through 43.

Sixty miles per hour equals 88 feet per second or 1 mile per minute. The proportion is: ft per sec: 88 ft per sec = mi per hr: 60 mi per hr.

39. How many feet per second does a car travel at a speed of 15 mph?   _____

40. Express in feet per second the speed of 45 mph.   _____

41. Express in feet per second the speed of 25 mph.   _____

42. Express in feet per second the speed of 40 mph.   _____

43. A track star can race at a speed of 33 feet per second. Express this speed in mph.   _____

    Note: Use this information for problems 44 through 46.

Proportion: mi per min: 1 mi per min = mi per hr: 60 mi per hr.

44. An airplane travels at a speed of 4 miles per minute. Express this speed in mph.   _____

45. A speedometer in a car is being checked. The car is driven at a uniform speed of 45 miles per hour for a measured mile. The time actually taken to travel the mile is 1 minute and 20 seconds.

    a. Is this the true speed?   a. _____

    b. Is the speedometer fast, slow, or correct?   b. _____

46. Over a measured mile, a uniform speedometer speed of 35 mph is kept. The time taken to drive the mile is 1½ minutes.

    a. Is the speedometer in error?   a. _____

    b. Is the speedometer fast, slow, or correct?   b. _____

47. How much longer does it take to travel 10 miles at 30 miles per hour than at 50 miles per hour?   _____

48. In 15 minutes, 10 miles can be traveled. How long does it take to travel 30 miles at the same average rate of speed?   _____

49. In 1 hour and 15 minutes, 48 miles are traveled. At the same average rate of speed, what distance is traveled in 4 hours and 45 minutes?   _____

50. A time trial for a racing car shows that on a ¾-mile track one lap is traveled in 38 seconds. At this rate, what time is made in a 100-mile race? Express the answer in hours, minutes, and seconds.   _____

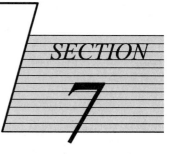

Formulas

Unit 32 FORMULAS FOR CIRCULAR MEASUREMENT

BASIC PRINCIPLES OF FORMULAS FOR CIRCULAR MEASUREMENT

Many parts that an automotive technician must measure are circular. The following formulas are necessary to solve problems involving circular measurement. A *formula* is a simplified way to express what must be done to solve a problem. Substitute the values that are known, then do the mathematical process to solve for the unknown value.

FORMULAS FOR CIRCULAR MEASUREMENT		
$C = \pi D$	$D = \dfrac{C}{\pi}$	C = circumference
$C = 2\pi R$	$R = \dfrac{C}{2\pi}$	D = diameter
$A = \pi R^2$	$D = \sqrt{\dfrac{A}{0.7854}}$	R = radius
	$R = \sqrt{\dfrac{A}{\pi}}$	A = area
		π = 3.1416

PRACTICAL PROBLEMS

1. The diameter of a circle is 12 inches.

   a. Find the circumference to the nearest thousandth inch. a. _____

   b. Find the radius. b. _____

2. The circumference of a circle is 50 centimeters. Express the following answers to the nearest thousandth centimeter.

   a. What is the measure of the diameter? a. _____

   b. What is the measure of the radius? b. _____

3. In a 10-inch radius circle

   a. Find the diameter. a. _____

   b. Find the circumference. b. _____

4. If the diameter of a piston is 78 millimeters (mm), find the area of the top of the piston. _____

5. On an engine equipped with an overhead camshaft, what is the length of the timing belt that is in contact with the camshaft sprocket? _____

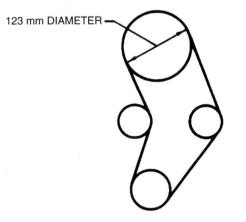

123 mm DIAMETER

6. At a certain speed, a car can stop in a distance of 49 meters. If the wheel and tire have a diameter of 61 centimeters, how many revolutions will the wheel have to make before the car comes to a complete stop? _____

The differential allows each rear wheel to turn independently. The need for a differential is easily understood, since in turning corners, one wheel must turn at a different rate of speed from the other. The *turning radius* of a car is the distance from the center of the rear axle to the pivot point. With a 15-foot turning radius, one wheel travels 7.3304 feet further than the other wheel in making a 90-degree turn.

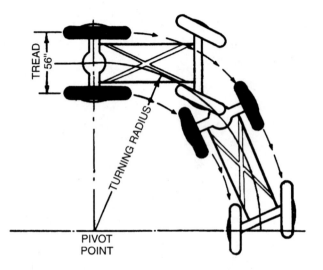

7. A 6.50 × 15-inch tire is used on the rear wheels. How many more turns will one rear wheel make than the other?   _____

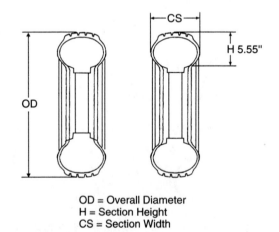

OD = Overall Diameter
H = Section Height
CS = Section Width

Hint:

    a. Find the overall diameter of the tire: D = 2 (Section Height) + Rim size.   _____

    b. Find the circumference of the tire: C = πD where π = 3.1416 (round to the nearest inch).   _____

c. Compare by division 7.3304 feet (the extra distance one wheel travels) with the circumference: _____

$$\frac{7.3304\ feet}{Circumference\ (in\ feet)} = Number\ of\ extra\ turns\ one\ wheel\ makes$$

Note: Use this diagram for problems 8 through 10.

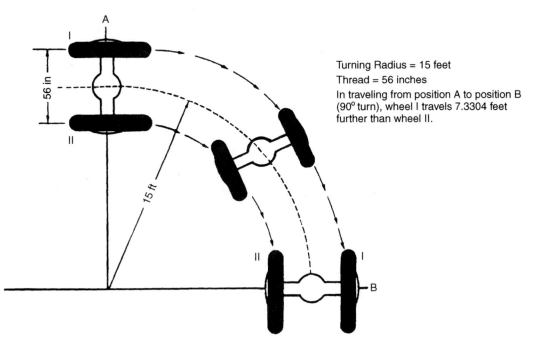

Turning Radius = 15 feet

Thread = 56 inches

In traveling from position A to position B (90° turn), wheel I travels 7.3304 feet further than wheel II.

8. A car makes a 90-degree turn with a turning radius of 15 feet. How many more turns will one rear wheel make if 235/60 R15 tires are used on a car? Express the answer to the nearest hundredth. (Note: Use the hint in problem 7.)   _____

9. The turning radius of a car is 15 feet. In order to make a U-turn, how many feet further will one wheel travel than another? (Hint: a U-turn is a 180-degree turn. It is the same as two 90-degree turns.)   _____

10. The turning radius of a car is 19 feet. How many feet further does one rear wheel travel than the other in making a 90-degree turn? The standard track or tread equals 56 inches, or 4⅔ feet.   _____

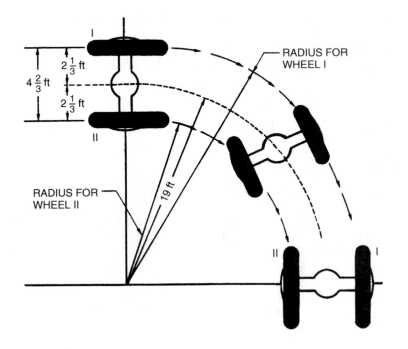

RADIUS FOR
WHEEL I

$2\frac{1}{3}$ ft

$4\frac{2}{3}$ ft

$2\frac{1}{3}$ ft

I

II

RADIUS FOR
WHEEL II

19 ft

II   I

Hint: 90 degrees is ¼ of a circle.

    a. To find the distance traveled by each wheel, use the formula:

       Distance = where $\dfrac{C}{4} = \pi D$.

       Distance wheel I   _____

       Distance wheel II   _____

    b. Subtract to find the difference of the distances.   _____

11. How wide should a road be to allow a U-turn if the turning radius is 19 feet?
     Express the answer in feet and inches. (Note: A U-turn is a 180-degree turn.
     The width of the road is the diameter for wheel I. Use the hint from problem
     10 to find the radius for wheel I.)   _____

12. Find the area of a circle with a diameter of 6 centimeters. Express the answer
     to the nearest thousandth square centimeter.   _____

13. The diameter of a circle is 0.50 meters. What is the area of the circle to the
     nearest thousandth square meter?   _____

14. Find the area of a ¼-inch diameter circle. Express the answer to the nearest thousandth square inch.   _____

15. What is the area of a circle with a radius of 3⅝ inches? Express the answer to the nearest thousandth square inch.   _____

16. Find the diameter of a piston with an area of 11.34 square inches. Express the answer to the nearest hundredth.   _____

17. The area of the top of an oil drum is 12.75 square feet. Find the diameter of the drum to the nearest hundredth foot.   _____

18. What diameter valve is needed to close an intake port with an area of 3¾ square inches? Express the answer to the nearest hundredth inch.   _____

19. A circle has an area of ⅞ square inch. What is the radius setting needed to lay out this circle? Express the answer to the nearest thousandth inch.   _____

20. On an overhead camshaft engine, the diameter of the base circle of the cam is 42 millimeters (mm). What is the distance the lobe of the camshaft travels while the valve is closed? (Hint: 242.2 degrees is ⅔ or 66.7% of a circle.)   _____

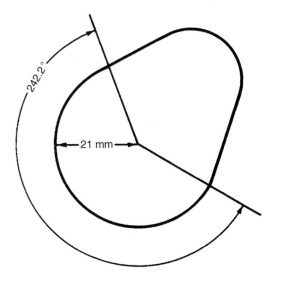

Unit 33 FORMULAS FOR EFFICIENCY

BASIC PRINCIPLES OF FORMULAS FOR EFFICIENCY

Study the following principles of efficiency. The formula for efficiency is:

$$\text{Efficiency} = \frac{\text{output}}{\text{input}}$$

where efficiency is expressed as a percent.

For *mechanical efficiency,* the output is expressed in foot-pounds of work obtained. The input is foot-pounds of work supplied.

In automobiles, mechanical efficiency compares the brake horsepower and the indicated horsepower. It is written

$$\text{Mechanical efficiency} = \frac{\text{brake horsepower (bhp)}}{\text{indicated horsepower (ihp)}}$$

Brake horsepower means the power delivered to the driving wheels. *Indicated horsepower* means the power delivered to the piston by the burning gas. For *thermal efficiency,* the output is the heat transformed into useful work. The input is the heat supplied.

In automobiles, because heat escapes by radiation and out the exhaust pipe, the thermal efficiency is low.

PRACTICAL PROBLEMS

1. In a car, the indicated horsepower developed by gasoline is 100. Only 74 horsepower is available for actual work. What is the mechanical efficiency? _____

2. In low gear, due to the friction of transmission gears, the brake horsepower is reduced to 83. The indicated horsepower is 100. What is the mechanical efficiency in this case? _____

3. The indicated horsepower is 65, but only 50 horsepower is delivered at the driving wheels. What is the mechanical efficiency? Express the answer to the nearest whole percent. _____

Note: Use this information for problems 4 through 6.

One gallon of gasoline equals 125,000 Btu; 1 Btu equals 778 foot-pounds of work. (Btu means British thermal unit.)

4. The number of British thermal units that are changed into useful work is 1,900. The number supplied is 3,500 Btu. What is the thermal efficiency to the nearest whole percent?   _____

5. Out of 50,000 Btu supplied, only 12,000 Btu are actually used in doing work. What is the thermal efficiency?   _____

6. One gallon of gasoline is used in pulling a load 10 miles. The force needed to pull the load is 500 pounds. What thermal efficiency does this represent? Express the answer to the nearest tenth percent.   _____

Unit 34 *TEMPERATURE*

BASIC PRINCIPLES OF TEMPERATURE

Two different scales are used in measuring temperatures. These scales are called *Fahrenheit (F)* and *Celsius (C)*. The automobile mechanic must be familiar with the process of changing from one scale to the other. The formula for expressing degrees Fahrenheit in degrees Celsius is

$$C = \frac{5}{9} (F - 32)$$

The formula for expressing degrees Celsius in degrees Fahrenheit is

$$F = \left(\frac{9}{5} \times C \right) + 32$$

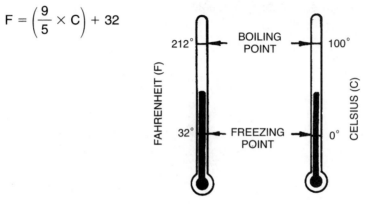

Example: Change 103 degrees Fahrenheit to Celsius.

$$\frac{5}{9} (F - 32)$$

Subtract 32 from 103: $103 - 32 = 71$

$$C = \frac{5}{9} \times \frac{71}{1} = \frac{355}{9} = 39\frac{4}{9} \text{ or } 39.44 \text{ degrees C}$$

Example: Change 20 degrees Celsius to Fahrenheit.

$$F = \left(\frac{9}{5} \times C \right) + 32$$

$$\frac{9}{5} \times \frac{20}{1} = 36 + 32 = 68 \text{ degrees F}$$

PRACTICAL PROBLEMS

1. Heat causes a change in materials, such as viscosity of oils or specific gravity of liquids. When this happens, values must be standardized at a definite temperature. This temperature is usually room temperature, 68 degrees Fahrenheit. What is the Celsius reading for this temperature?   _____

2. Alcohol boils at 179 degrees Fahrenheit. What is the Celsius reading to the nearest tenth degree?   _____

Express these melting point temperatures to the nearest tenth degree.
   Note: Use this chart for problems 3 through 7.

Problem	Material	Degrees Fahrenheit	Degrees Celsius
3.	Cast iron	?	1,260
4.	Aluminum	1,200	?
5.	Lead	?	327
6.	Copper	1,940	?
7.	Chromium	2,740	?

3. _____

4. _____

5. _____

6. _____

7. _____

Unit 35 CYLINDRICAL VOLUME MEASUREMENT

BASIC PRINCIPLES OF CYLINDRICAL VOLUME MEASUREMENT

Refer to Unit 32 before attempting any problems in this unit.

The basic formula for cylindrical measurement is:

$$\text{Volume} = \pi r^2 \text{ (Area of a circle)} \times \text{Length of cylinder}$$

Volume is expressed in cubic units.

In automotive work, the diameter is frequently referred to as the *bore*.

The *capacity,* in gallons, of cylindrical tanks can be found by using the formula:

$$C = \frac{0.7854D^2L}{231 \text{ cu in per gal}}$$

where: C = capacity in gallons
D = diameter of tank in inches
L = length of tank in inches
231 cubic inches = 1 gallon

Total *piston displacement (PD)* is the volume (number of cubic units) displaced as the piston moves from BDC to TDC. The formula is written

$$PD = 0.7854D^2LN$$

where: D = diameter of bore of cylinder
L = length of stroke (inches)
N = number of cylinders

PRACTICAL PROBLEMS

1. The cylindrical gasoline tank on a tractor is 10 inches in diameter and 30 inches long. How many gallons will the tank hold? _____

2. A storage tank in a filling station measures 8 feet in length and 4 feet in diameter. What is the capacity of the tank to the nearest hundredth gallon? _____

3. A drum of alcohol measures 4 feet in height and 25 inches in diameter. Is this a 25-, 50-, 100-, 200-, or 500-gallon drum?   _____

4. It is necessary to get a 50-gallon tank in a space 3 feet long. What size diameter is needed? Express the answer to the nearest tenth.   _____

5. A cylindrical tank is 25 inches in diameter and holds 500 gallons. What is the length of the tank to the nearest inch?   _____

6. A 6-cylinder engine has a 3.750-inch bore and a $3\frac{3}{4}$-inch stroke. Find the total piston displacement to the nearest thousandth cubic inch.   _____

7. What is the total piston displacement on an 8-cylinder engine with a $3\frac{1}{2}$-inch diameter and $4\frac{3}{8}$-inch stroke? Express the answer to the nearest thousandth cubic inch.   _____

8. In an engine, the piston diameter is $3\frac{1}{4}$ inches, the stroke is $4\frac{7}{8}$ inches, and the piston displacement is 323.5 cubic inches. Find the number of cylinders.   _____

9. A 6-cylinder engine with a piston displacement of 235.5 cubic inches has a $3\frac{15}{16}$-inch stroke. Find the piston diameter of the engine to the nearest hundredth inch.   _____

10. A 4-cylinder Honda has a bore of 80 millimeters and a stroke of 91 millimeters. Find the piston displacement in cubic centimeters. (Hint: First change millimeters to centimeters.)   _____

11. A Toyota 4-cylinder has a displacement of 1995 cubic centimeters. If the bore is 3.31 inches and the stroke is 3.54 inches, find the piston displacement in cubic inches.   _____

12. A Volvo diesel V-6 has a bore of 7.65 centimeters and a stroke of 8.3 centimeters. Find the displacement in cubic centimeters and liters.   _____

13. The same year Volvo made a gasoline fuel injected V-6 that had a bore of 9.09 centimeters and a stroke of 7.26 centimeters. Which engine had the larger displacement and by how many cubic centimeters?   _____

14. A BMW 4-cylinder 318i has an 8.9-centimeter bore and a 7-centimeter stroke. Find the displacement in cubic centimeters and liters.   _____

15. Find the displacement of an 8-cylinder engine with a bore of 3.622 inches and a stroke of 3.307 inches.   _____

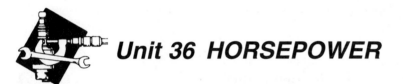

Unit 36 HORSEPOWER

BASIC PRINCIPLES OF HORSEPOWER

The horsepower for gasoline, steam, diesel, or compressed air engines can be found using the formula:

$$hp = \frac{PLAN}{33,000}$$

where: P = average pressure (pounds per square inch)
L = length of stroke (feet)
A = area of piston (square inches)
N = number of power strokes per minute

SAE horsepower is a method of estimating the horsepower. It is used as a rating. SAE horsepower is sometimes used to determine the amount to be paid for taxation. The formula for SAE horsepower is:

$$hp = \frac{D^2N}{2.5}$$

where: D^2 = cylinder bore squared
N = number of cylinders

PRACTICAL PROBLEMS

1. An automobile engine has a bore of 3.5000 inches, a stroke of 2¾ inches, average pressure of 125 pounds per square inch, and 3,000 power strokes per minute. Find the horsepower of the engine to the nearest tenth. _____

2. An 8-cylinder engine has a crankshaft speed of 1,500 rpm, a bore of 3.750 inches, a stroke of 4 inches, and average pressure of 100 pounds per square inch. Find, to the nearest tenth, the horsepower of the engine. _____

3. What effect does doubling the speed of the crankshaft have on the horsepower? _____

4. A 6-cylinder car has a bore of 3.875 inches. What is the horsepower rating to the nearest tenth? _____

5. What is the SAE horsepower of an 8-cylinder engine with a bore of 3.25 inches?   _____

6. In a car, the horsepower rating is 21.6, and the bore is 3 inches. How many cylinders are there?   _____

    Note: $N = \dfrac{hp \times 2.5}{D^2}$

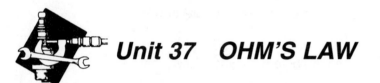

Unit 37 OHM'S LAW

BASIC PRINCIPLES OF OHM'S LAW

Ohm's Law states the relationship of pressure, current flow, and resistance in an electrical circuit. Understanding this relationship helps the automotive technician diagnose problems in the automobile electrical system. Ohm's Law is expressed in these three statements:

1. If voltage is increased, current flow is increased.

2. If voltage is decreased, current flow is decreased.

3. If resistance is increased, current flow is decreased.

The formula can be written three ways:

$$E = I \times R \qquad I = \frac{E}{R} \qquad R = \frac{E}{I}$$

where: E = voltage (electromotive force)
           I = current in amperes (inductance)
           R = resistance in ohms

These three forms of Ohm's Law are simply three ways of expressing the same principle.

PRACTICAL PROBLEMS

1. The lighting system of an automobile draws $5\frac{1}{2}$ amperes at a battery voltage of $11\frac{3}{4}$. Find the resistance of the lighting system to the nearest thousandth ohm. _____

2. While starting the motor, 250 amperes are flowing. With a 12-volt battery, what is the resistance of the starting motor? _____

3. The blower motor in a car heater has a resistance of 9 ohms. What current is flowing from the 12-volt battery? Express the answer to the nearest thousandth. _____

4. An automobile horn requires only 0.4 ampere. The resistance is 30 ohms. Find the voltage. _____

5. Find the current flow (amperes) in a headlight circuit with 1.6 ohms resistance. The battery voltage is measured at 12.6 volts.   _____

6. How many amperes are flowing in an ignition circuit that has 3 ohms total resistance and a battery voltage of 12.6 volts?   _____

Note: In a series circuit, all of the resistances are added together and then substituted in Ohm's Law together to equal total resistance (R). In a parallel circuit, a formula for resistance must be used.

$$\text{Total resistance (R)} = \frac{1}{\frac{1}{r_1} + \frac{1}{r_2} + \frac{1}{r_3}} \text{ or } \frac{r_1 \times r_2}{r_1 + r_2}$$

Some technicians prefer to use the second formula even if they have more than two resistances. First solve for two resistances, which becomes r_1, and then for two more resistances, which becomes r_2. Then solve for total resistance.

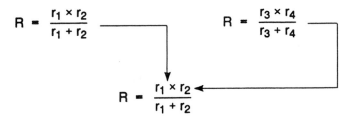

It should be noted that in a parallel circuit, the total resistance is always less than the lowest resistance in the circuit. Refer to Unit 13 if you need to review addition of fractions.

7. In the series circuit below, solve for current flow.   _____

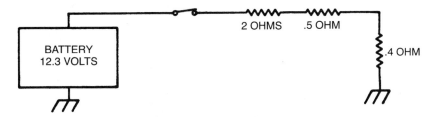

8. In the parallel circuit below, solve for total resistance in the circuit.   _____

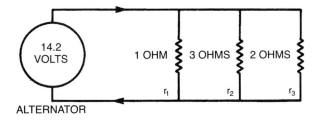

Note: Use this diagram for problems 9 through 11.

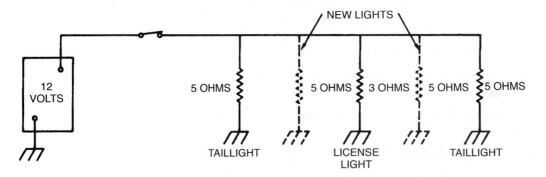

9. A mechanic wants to add two taillights to a car with a taillight circuit like that in the diagram. What will be the total resistance in the new circuit if the mechanic uses bulbs of the same resistance?   _____

10. What is the total resistance in the circuit before the new lights are added in problem 9?   _____

11. Find the total current flow in the diagram. The rule to find current flow (amperes) is: The total current flow in a parallel circuit is the sum of the individual circuits.   _____

Hint: Use $I = \dfrac{E}{R}$ for each blank, then add each answer.

12. Find the total resistance in the figure below.   _____

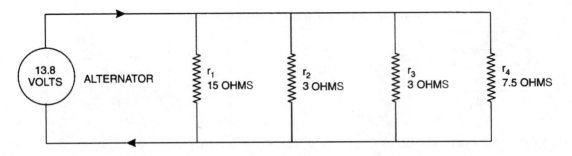

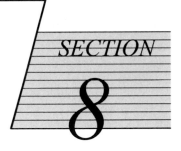

Graphs

Unit 38 USE OF GRAPHS

BASIC PRINCIPLES OF GRAPHS

A graph is a diagram that allow you to compare data on two or more variables at the same time. There are three types of graphs used in this unit—bar, line, and circle graphs. The circle graph is sometimes referred to as a pie chart.

PRACTICAL PROBLEMS

This bar graph shows the average braking distance for a car. The *braking distance* is measured from the time the brakes are applied until the time the car comes to a full stop. The car is a medium size and weight and is stopping on dry pavement.

Note: Use this graph for problems 1 through 3.

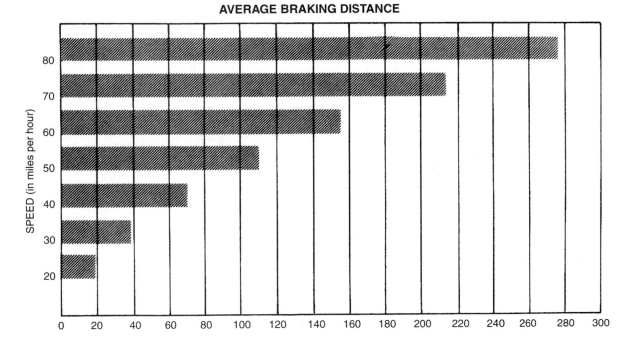

AVERAGE BRAKING DISTANCE

1. At a speed of 50 mph, what is the braking distance?   _____

2. If the braking distance is about 155 feet, what is the speed of the car?   _____

3. How many more feet does it take to stop a car at 70 mph than at 30 mph?   _____

4. Using this graph, find the number of heat units used in getting a substance to a temperature of 105°F.   _____

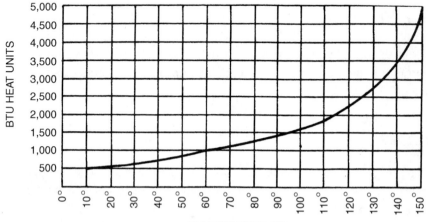

HEAT UNITS USED TO REACH VARIOUS TEMPERATURES

This graph shows how the average mechanic's salary changes throughout the year.

   Note: Use this graph for problems 5 through 8.

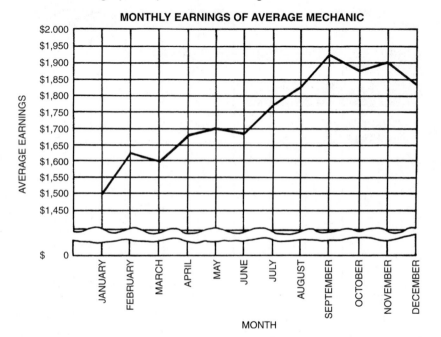

MONTHLY EARNINGS OF AVERAGE MECHANIC

5. How much does the average mechanic earn in May?   _____

6. What are the earnings in July?   _____

7. What is the least amount the average mechanic makes during the year?   _____

8. During which month is the largest amount earned?   _____

9. From this circle graph, find the percent of an average person's life that is spent at work.   _____

AVERAGE PERSON'S LIFE

10. A family budgets $1,500 per month as follows:

Insurance	$ 25.00
Food	$350.00
Miscellaneous	$100.00
Rent and utilities	$550.00
Savings	$ 95.00
Clothes	$100.00
Car payment	$225.00
Charity	$ 55.00

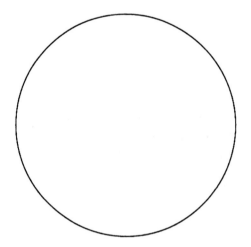

Make a circle graph of the budget.

11.   Plot a line graph of a parts manufacturer's output for the first two years. Use a solid line (or blue pencil) for the first year and a dashed line (or red pencil) for the second year.

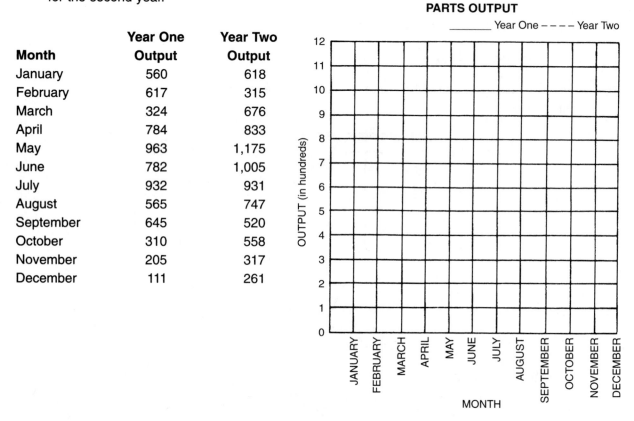

Month	Year One Output	Year Two Output
January	560	618
February	617	315
March	324	676
April	784	833
May	963	1,175
June	782	1,005
July	932	931
August	565	747
September	645	520
October	310	558
November	205	317
December	111	261

12.   Before the brakes are applied, the mind and body must react. The mind must realize that there is a need to apply the brakes, and the foot must be moved and placed on the brakes. This is called the *reaction time.* The distance traveled during the reaction time changes with speed. From this table, make a bar graph of the speed and the reaction distance.

SPEED (in mph)	DISTANCE (in feet)
20	21
30	31
40	41
50	51
60	62
70	72
80	82

AVERAGE REACTION DISTANCE

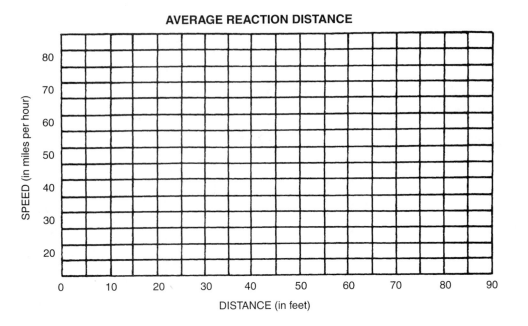

This graph shows the increased efficiency of an automobile engine equipped with a high-compression head.

Note: Use this graph for problems 13 through 15.

POWER

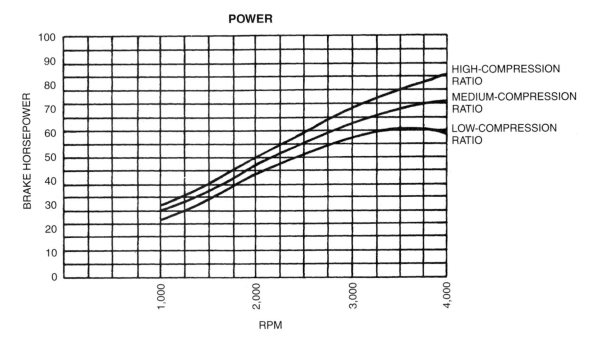

13. What brake horsepower is developed by the engine having a low-compression ratio when turning over at 3,500 rpm?   _____

14. Two engines, one equipped with a low-compression head and the other with a high-compression head, rotate at 2,500 rpm. What is the difference in generated brake horsepower between the two engines? _____

15. How many rpm will a medium-compression head motor be making in order to generate 60 brake horsepower? _____

Note: Use these graphs for problems 16 through 19.

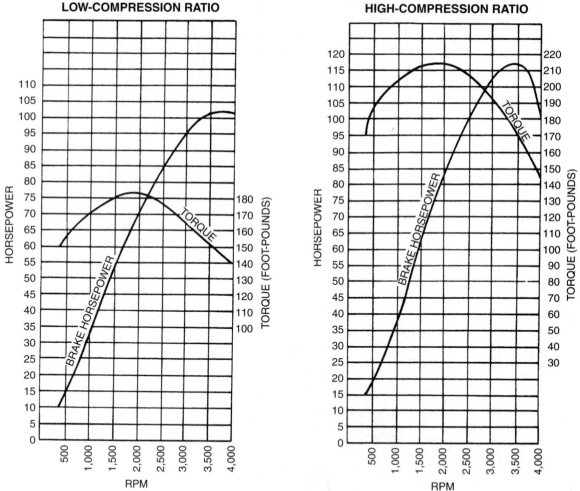

16. At what rpm of a low-compression ratio is maximum torque attained? _____

17. For a high-compression ratio, at what rpm is maximum horsepower attained? _____

18. Maximum horsepower of a low-compression ratio occurs at what speed?   _____

19. At what rpm of a high-compression ratio does maximum torque occur?   _____

Note: Use this graph for problems 20 and 21.

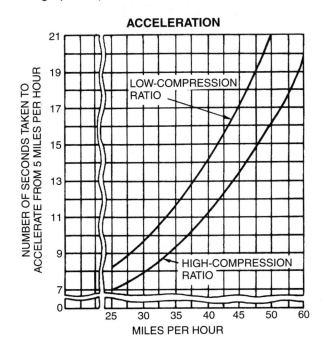

20. A car starting at 5 miles per hour accelerates for 17 seconds. What is the fastest possible speed for a car having a high-compression head?   _____

21. A car with a low-compression head accelerates from 30 mph to 40 mph. How many seconds are needed to do this?   _____

Note: Use this graph for problems 22 through 25.

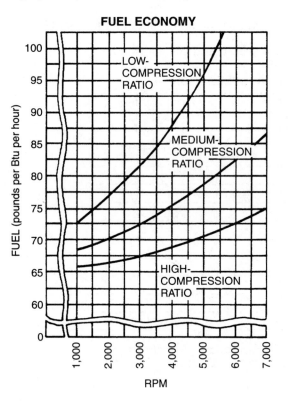

22. At 3,000 rpm, the car that uses the most fuel has what type of compression ratio?   _____

23. At 3,000 rpm, what is the maximum fuel a low-compression ratio car uses?   _____

24. The fuel for a car with a medium-compression head is 70 pounds per Btu per hour. How fast does the automobile engine rotate in rpm?   _____

25. A car having a low-compression head and one having a high-compression head both rotate at 2,750 rpm. What is the difference in fuel consumption in the two cars?   _____

Note: Use this graph for problems 26 through 28.

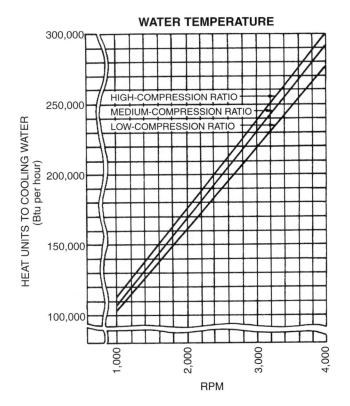

26. A low-compression motor is running at 2,750 rpm. How many heat units per hour are returning to the cooling water?   _____

27. A medium-compression ratio motor returns 250,000 heat units per hour to the cooling water. How fast is the motor rotating?   _____

28. Cars having a low-compression ratio and ones having a high-compression ratio rotate at 3,250 rpm. What is the difference between the amounts of heat returned to the cooling water by the cars?   _____

Note: Use this graph for problems 29 through 31.

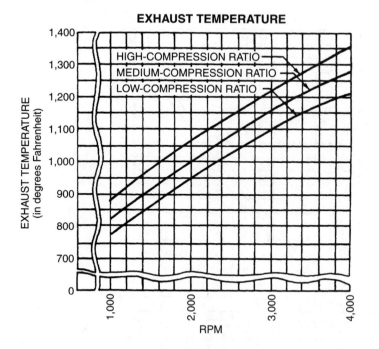

EXHAUST TEMPERATURE

29. At 2,250 rpm, what is the exhaust temperature of the engine having a medium-compression ratio?   _____

30. At 1,150°F exhaust temperature, how fast is an engine with a high-compression ratio rotating?   _____

31. At 3,000 rpm, what is the difference in exhaust temperatures between a high-compression and a low-compression ratio car?   _____

Emission control on modern automobiles is one of the biggest problems that the automotive engineer faces. After the car is sold, it is up to the automotive technician to maintain the emission controls at "like-new" performance levels. The "Big Three" of pollution are hydrocarbons (HC), carbon monoxide (CO), and oxides of nitrogen (NO$_x$). This graph compares the total percent reduction for these emissions over a five-year period. **Note:** Use this graph for problems 32 and 33.

TOTAL REDUCTION IN EMMISSIONS OVER A 5-YEAR PERIOD

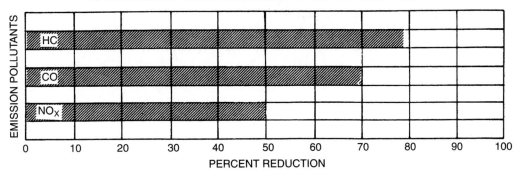

32. Which of these three pollutants have the engineers had the most success in controlling?   _____

33. Which of the three is the most difficult to bring under control?   _____

The mechanic can help reduce emissions by correct tune-up procedures, especially idle mixture adjustments. In these two graphs, the effect on emissions of too lean a mixture or too rich a mixture is shown.

A mechanic can use these graphs and an infrared analyzer to correct idle settings. As the idle screw is adjusted toward lean, both meters decrease. When the hydrocarbons increase, the mechanic has a lean idle miss. The mixture is then enriched slightly. **Note:** Use these graphs for problems 34 through 36.

EFFECT OF IDLE MIXTURE ADJUSTMENTS ON EMISSIONS

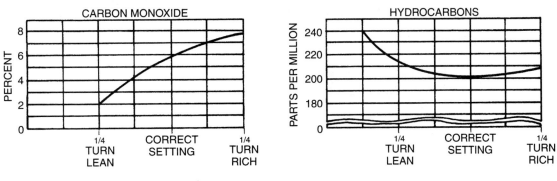

34. When the mixture screw is turned from the correct setting to ¼ turn rich, what happens to carbon monoxide and hydrocarbon levels?   _____

35. When the mixture screw is turned from the correct setting to ¼ turn lean, what happens to the carbon monoxide level?   _____

36. What happens to the hydrocarbon level when the setting is turned from the correct setting to ¼ turn lean?   _____

Emission levels are affected when one cylinder is not firing. These two graphs show the results.

Note: Use these graphs for problems 37 and 38.

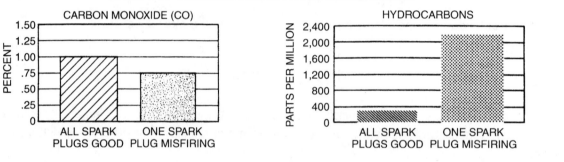

EFFECT OF SHORTED SPARK PLUGS ON EMISSIONS

37. When one plug misfires, about what percent does the CO level decrease?   _____ %

38. How much does the hydrocarbon level rise when one plug misfires? Express the answer in parts per million.   _____ PPM

The data in this chart was obtained from an actual test of an engine with the crankshaft speed held constant at 3,000 rpm while the spark settings were varied. **Note:** Use this chart for problem 39.

Total Spark Advance (in degrees)	Brake Horsepower	Exhaust Valve Temperatures (in degrees Fahrenheit)
10	58.5	1,350
20	77.0	1,280
30	84.0	1,255
40	86.5	1,255
50	84.0	1,300

39. Using the chart, make a graph with two curves. With a solid line show how changes in spark setting affect the brake horsepower. With a dotted line show how changes in the spark settings affect the exhaust valve temperature.

    Note: Use the completed graph for problems 40 through 44.

EFFECT OF SPARK ADVANCE ON BRAKE HORSEPOWER AND EXHAUST TEMPERATURE

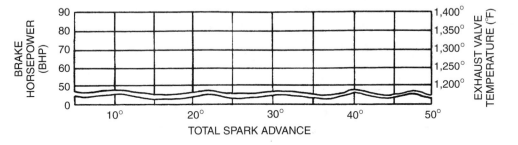

40. What spark setting gives the greatest horsepower?   _____

41. When the spark is advanced from 10° to 40°, what is the effect on the horsepower?   _____

42. What is the effect on the exhaust valve temperature when the spark is advanced from 10° to 30°?   _____

43. What effect on the brake horsepower does a spark advance of 10° have?   _____

44. When the spark is advanced to 10°, what is the effect on the exhaust valve temperature?   _____

A *flow test* is sometimes needed to find out if a radiator is clogged. In a flow test, the quantity of water that will flow through a radiator (by gravity) in a given time is measured. Certain standards are set by the manufacturer. If less water than the standard flows through the radiator, it is clogged. This chart shows the standards for certain model cars. **Note:** Use this chart for problem 45.

Car Model	Flow Test Standard
Buick	27 gallons per minute
Chevrolet	19 1/2 gallons per minute
Ford	42 1/2 gallons per minute
Plymouth	17 1/2 gallons per minute

45.   In these four graphs, plot the time (in seconds) against the flow of water (in gallons). Use the standards set for each car.

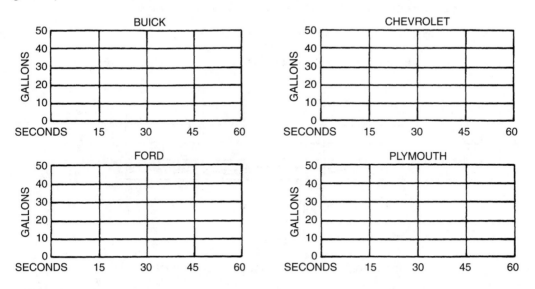

Note: Use these graphs for problems 46 through 49.

46.   In a radiator test on a Buick, a container is placed under the bottom radiator outlet for 10 seconds, and 4 gallons of water are collected. Determine if this radiator test does or does not indicate clogging.   _____

47.   A Chevrolet is given the radiator test, and 26 quarts of water are collected in 20 seconds. Does this test indicate clogging?   _____

48.   For a Plymouth radiator test, water is going to be collected for 30 seconds. How many quarts should a container hold?   _____

49.   A Ford is to be given the radiator test. A 5-gallon can is the largest container available. For how many seconds may the test be run, assuming the radiator is not clogged?   _____

CALCULATING CHART

For Increasing the Protection of Ethylene Glycol Antifreeze Solution

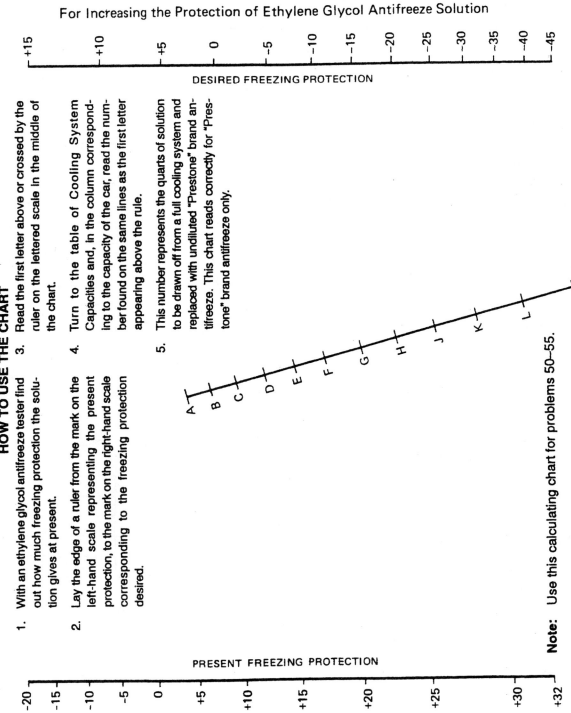

DESIRED FREEZING PROTECTION

+15 +10 +5 0 −5 −10 −15 −20 −25 −30 −35 −40 −45

PRESENT FREEZING PROTECTION

−20 −15 −10 −5 0 +5 +10 +15 +20 +25 +30 +32

HOW TO USE THE CHART

1. With an ethylene glycol antifreeze tester find out how much freezing protection the solution gives at present.

2. Lay the edge of a ruler from the mark on the left-hand scale representing the present protection, to the mark on the right-hand scale corresponding to the freezing protection desired.

3. Read the first letter above or crossed by the ruler on the lettered scale in the middle of the chart.

4. Turn to the table of Cooling System Capacities and, in the column corresponding to the capacity of the car, read the number found on the same lines as the first letter appearing above the rule.

5. This number represents the quarts of solution to be drawn off from a full cooling system and replaced with undiluted "Prestone" brand antifreeze. This chart reads correctly for "Prestone" brand antifreeze only.

Note: Use this calculating chart for problems 50–55.

This is a list of cooling system capacities and quarts of ethylene glycol antifreeze required. This list is to be used in connection with the Calculating Chart.

Quarts to Replace	Cooling System Capacity					
	2 gal.	3 gal.	4 gal.	5 gal.	6 gal.	7 gal.
A		1	1	1	1	1
B	1	1	2	2	2	3
C	1	2	2	3	4	4
D	2	2	3	4	5	6
E	2	3	4	5	6	7
F	2	4	5	6	7	8
G	3	4	6	7	8	10
H	3	5	6	8	10	11
J	4	5	7	9	11	13
K	4	6	8	10	12	14
L	4	7	9	11	13	15

Note: Use this chart for problems 50 through 55.

Find the number of quarts of ethylene glycol antifreeze needed for the conditions stated in each problem.

Problem	Capacity of System	Present Protection	Desired Protection	Quarts of Ethylene Glycol Required
50.	2 gal.	+20	-20	
51.	3 gal.	+32	-20	
52.	4 gal.	+10	-10	
53.	5 gal.	+20	-10	
54.	6 gal.	+15	-15	
55.	7 gal.	+10	-20	

Invoices

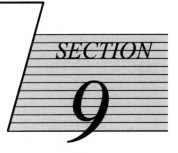

Unit 39 REPAIR ORDERS AND SHOP TICKETS

BASIC PRINCIPLES OF REPAIR ORDERS AND SHOP TICKETS

Many times a mechanic is asked or required by the shop owner to fill out shop tickets or invoices for the customer. If the mechanic makes a mistake in favor of the customer (charges too little), there may be trouble with the employer. After all, the owner is in business to make a profit. On the other hand, a mistake in favor of the employer is just as serious. Although the mistake may be unintentional, the customer may not think so. Sometimes, if not handled properly, such a mistake may cause the shop to lose a good customer.

Many states have laws to protect the consumer. The mechanic is required to give an accurate estimate, within certain limits, on each job. You should familiarize yourself with your local laws.

Apply these principles to the automotive field by solving the review problems that follow.

PRACTICAL PROBLEMS

Following each problem is a shop ticket/invoice to provide practice that will be helpful. Complete the shop tickets and use them to solve the problems.

Problem 1

A major tune-up on an 8-cylinder automobile requires the following parts and charges:

Item	Quantity Needed	Cost
8 Spark plugs	1 set	$23.84
Spark plug wires	1 set	$55.83
Distributor cap	1	$18.40
Rotor	1	$6.80 each
Carburetor kit	1	$33.15
Rebuilding carburetor	Flat rate	$45.00
Labor	Flat rate	$42.00
5% sales tax on parts only		

1. Complete the invoice for this job.

2. The mechanic receives 50% of the labor charge. How much is received for this job?   _____

3. If the job requires 3.2 hours to complete, what does the mechanic average per hour? Express the answer to the nearest cent.   _____

4. The shop owner pays $1.76 for each spark plug. The owner pays 40% off list price for all other materials. What is the shop owner's profit on the parts?   _____

Problem 2

An air-conditioning repair job requires the following parts:

> 1 seal kit @ $15.80 each
> 3 pounds refrigerant (R-12) @ $2.95 per pound
> 2 ounces oil @ $1.08 per ounce
> 42 inches $^{13}\!/_{32}$-inch hose @ $4.98 per foot
> 2 hose clamps @ $0.95 each

Labor charge at $42.00 per hour is as follows:

Job	Time Required
Seal compressor	1.6 hours
Add oil	0.2 hours
Replace hose and clamps	0.5 hours
Charge with refrigerant	1.1 hours

Include 6% sales tax on parts only.

1. Write up the work order for this job.

2. A mechanic's commission is 50% of this labor charge. What is the average pay per hour for this job?   _____

3. The mechanic is guaranteed $340.00 per week for 40 hours of work. What is the guaranteed hourly pay?   _____

4. All parts used on this job carry a 45% discount off list price. The owner receives 50% of the labor charge. What profit does the owner of the garage make on this job for parts and labor combined?   _____

PROBLEM 1– REPAIR ORDERS AND SHOP TICKETS

MIDTOWN GARAGE

212 NORTHERN BLVD.
BROWNVILLE, NEW YORK 13000
TELEPHONE 853-4008

NAME

ADDRESS

CITY   PHONE

MAKE	MODEL	YEAR	SER. NO.
			MTR. NO.

LICENSE NO.

SPEEDOMETER

RECEIVED	A. M.	DATE
	P. M.	Customer Order No.
PROMISED	A. M.	
	P. M.	
TERMS		Order Written By

OPER. NO.

REPAIR ORDER INSTRUCTIONS

LUBRI-CATE   CHANGE OIL   FLUSH TRANS   FLUSH DIFF   WASH   POLISH

LABOR CHARGE

I hereby authorize the above repair work to be done along with the necessary material, and hereby grant you and/or your employees permission to operate the car or truck herein described on streets, highways or elsewhere for the purpose of testing and/or inspection. An express mechanic's lien is hereby acknowledged on above car or truck to secure the amount of repairs thereto

X

NOT RESPONSIBLE FOR LOSS OR DAMAGE TO CARS OR ARTICLES LEFT IN CARS IN CASE OF FIRE, THEFT OR ANY OTHER CAUSE BEYOND OUR CONTROL.

PHONE WHEN READY:

F. S.

TOTAL LABOR	
TOTAL PARTS	
ACCESSORIES	
GAS, OIL & GREASE	
OUTSIDE REPAIRS	
TAX	
TOTAL AMOUNT	

GAS OIL and GREASE	PRICE
GALS. GAS @	
QTS. OIL @	
LBS GREASE @	
TOTAL GAS OIL AND GREASE	

MATERIAL USED

QUAN.	PART NO.	DESCRIPTION	PRICE

OUTSIDE REPAIRS

BROUGHT FORWARD

TOTAL PARTS

ACCESSORIES

QUAN.	ACCES. NO.	ACCESSORIES	PRICE

TOTAL ACCESSORIES

GRO-1 ®

PROBLEM 2 – REPAIR ORDERS AND SHOP TICKETS

MIDTOWN GARAGE

212 NORTHERN BLVD.
BROWNVILLE, NEW YORK 13000
TELEPHONE 853-4008

NAME

ADDRESS

CITY   PHONE

MAKE   MODEL   YEAR   SER. NO.
                                     MTR. NO.

OPER. NO.

RECEIVED   A.M.   DATE
            P.M.
PROMISED   A.M.   CUSTOMER ORDER NO.
            P.M.
TERMS   ORDER WRITTEN BY

LICENSE NO.

SPEEDOMETER

REPAIR ORDER INSTRUCTIONS

LUBRI- CHANGE FLUSH   FLUSH   WASH   POLISH
CATE   OIL   TRANS. DIFF.

LABOR
CHARGE

I hereby authorize the above repair work to be done along with the necessary material, and hereby grant you and/or your employes permission to operate the car or truck herein described on streets, highways or elsewhere for the purpose of testing and/or inspection. An express mechanic's lien is hereby acknowledged on above car or truck to secure the amount of repairs thereto.

X

NOT RESPONSIBLE FOR LOSS OR DAMAGE TO CARS OR ARTICLES LEFT IN CARS IN CASE OF FIRE, THEFT OR ANY OTHER CAUSE BEYOND OUR CONTROL.

GAS OIL AND GREASE

GALS. GAS ⬤
QTS. OIL ⬤
LBS. GREASE ⬤

TOTAL GAS. OIL AND GREASE

PRICE

F. S.

PHONE WHEN READY: ◯

TOTAL LABOR

TOTAL PARTS

ACCESSORIES

GAS. OIL & GREASE

OUTSIDE REPAIRS

TAX

TOTAL AMOUNT

MATERIAL USED

QUAN.	PART NO.	DESCRIPTION	PRICE

OUTSIDE REPAIRS

BROUGHT FORWARD

TOTAL PARTS

QUAN	ACCES. NO.	ACCESSORIES	PRICE

TOTAL ACCESSORIES

GRO.-1 ⬤

Problem 3

While a cooling system is being cleaned it is learned that the radiator is clogged and must be cleaned. The car owner asks that all the hoses, the thermostat, and the pressure cap be replaced. The job requires the following parts:

Item	Quantity	Cost
$\frac{5}{8}$-inch hose	30 inches	$1.25 per foot
$\frac{3}{4}$-inch hose	24 inches	$1.35 per foot
$\frac{5}{8}$-inch hose clamps	2	$0.84 each
$\frac{3}{4}$-inch hose clamps	2	$0.84 each
Thermostat with gasket	1	$5.65 each
Pressure cap	1	$4.50 each
Upper hose	1	$11.44 each
Lower hose	1	$16.53 each
Hose clamps	4	$0.98 each

Labor charges are as follows:

Remove, clean, and replace radiator	$39.50
Reverse flush block	$20.00
Replace heater hoses	$10.00
Replace thermostat and cap	N/C

Include 7% sales tax on parts and labor.

1. Complete the job ticket for this job.

2. The mechanic receives 60% of the labor charges. What is the pay for this job? _____

3. The total job takes 3 hours. What is the mechanic's average hourly earning, to the nearest cent? _____

4. The garage owner receives 40% discount for parts. How much is paid for the parts for this job, to the nearest cent? _____

5. After labor and parts cost, how much does the garage owner make on this job? _____

6. What percent of the profit that the owner receives is due to parts sales? _____

PROBLEM 3 – REPAIR ORDERS AND SHOP TICKETS

MIDTOWN GARAGE

212 NORTHERN BLVD.
BROWNVILLE, NEW YORK 13000
TELEPHONE 853-4008

RECEIVED A.M. P.M. | DATE
PROMISED A.M. P.M. | CUSTOMER ORDER NO.
TERMS | ORDER WRITTEN BY

NAME
ADDRESS
CITY | PHONE
MAKE | MODEL | YEAR | SER. NO. | LICENSE NO.
| | | MTR. NO. | SPEEDOMETER

OPER. NO.

REPAIR ORDER INSTRUCTIONS

LUBRI-CATE | CHANGE OIL | FLUSH TRANS. | FLUSH DIFF. | WASH | POLISH

LABOR CHARGE

I hereby authorize the above repair work to be done along with the necessary material, and hereby grant you and/or your employees permission to operate the car or truck described on streets, highways or elsewhere for the purpose of testing and/or inspection. An express mechanic's lien is hereby acknowledged on above car or truck to secure the amount of repairs thereto.

X

NOT RESPONSIBLE FOR LOSS OR DAMAGE TO CARS OR ARTICLES LEFT IN CARS IN CASE OF FIRE, THEFT OR ANY OTHER CAUSE BEYOND OUR CONTROL

GAS OIL AND GREASE | PRICE
GALS. GAS @
QTS. OIL @
LBS. GREASE @
TOTAL GAS, OIL AND GREASE

F. S.

PHONE WHEN READY:

TOTAL LABOR
TOTAL PARTS
ACCESSORIES
GAS, OIL & GREASE
OUTSIDE REPAIRS
TAX
TOTAL AMOUNT

MATERIAL USED

QUAN. | PART NO. | DESCRIPTION | PRICE

OUTSIDE REPAIRS

BROUGHT FORWARD

TOTAL PARTS

ACCESSORIES

QUAN. | ACCES. NO. | ACCESSORIES | PRICE
| | |
| | |

TOTAL ACCESSORIES

GRO. 1 ®

Problem 4

A car breaks down 12 miles from a garage. Towing service is $45.00 for a 3-mile radius and $3.50 per mile thereafter. The towing charge is based on one-way mileage. Sales tax of 5% is added to the charge.

1. Write up the shop ticket for the towing service.

2. The mechanic averages 15 miles per hour for the round trip. How long is the mechanic away from the shop?   _____

3. What does the shop owner average per hour for the call, not considering operating expenses or tax? Express the answer to the nearest cent.   _____

4. The mechanic receives 50% of the shop charges as salary. What does the mechanic earn for the trip?   _____

5. What is the mechanic's average pay per hour for the trip?   _____

6. The shop owner figures the expenses of operating the truck in terms of percents of the towing charges.

Expense	Percent of Towing Charge
Mechanic (driver)	50%
Gas and oil	4%
Insurance	4%
Depreciation	5%
Tires and miscellaneous	3%
Shop overhead	10%

What is the shop owner's profit for the trip? Tax is not part of the profit.   _____

PROBLEM 4 – REPAIR ORDERS AND SHOP TICKETS

MIDTOWN GARAGE

212 NORTHERN BLVD.
BROWNVILLE, NEW YORK 13000
TELEPHONE 853-4008

NAME

ADDRESS

CITY   PHONE

MAKE   MODEL   YEAR   SER. NO.   LICENSE NO.   SPEEDOMETER
                         MTR. NO.

RECEIVED   A.M. P.M.   DATE
PROMISED   A.M. P.M.   CUSTOMER ORDER NO.
TERMS   ORDER WRITTEN BY

OPER. NO.

REPAIR ORDER INSTRUCTIONS   LABOR CHARGE

LUBRICATE   CHANGE OIL   FLUSH TRANS.   FLUSH DIFF.   WASH   POLISH

I hereby authorize the above repair work to be done along with the necessary material, and hereby grant you and/or your employees permission to operate the car or truck herein described on streets, highways or elsewhere for the purpose of testing and/or inspection. An express mechanic's lien is hereby acknowledged on above car or truck to secure the amount of repairs thereto

X

NOT RESPONSIBLE FOR LOSS OR DAMAGE TO CARS OR ARTICLES LEFT IN CARS IN CASE OF FIRE, THEFT OR ANY OTHER CAUSE BEYOND OUR CONTROL

F. S.

GAS OIL AND GREASE		PRICE
GALS. GAS @		
QTS. OIL @		
LBS. GREASE @		
TOTAL GAS. OIL AND GREASE		

PHONE WHEN READY: ○

TOTAL LABOR
TOTAL PARTS
ACCESSORIES
GAS. OIL & GREASE
OUTSIDE REPAIRS
TAX
TOTAL AMOUNT

MATERIAL USED

QUAN.	PART NO.	DESCRIPTION	PRICE

OUTSIDE REPAIRS

BROUGHT FORWARD

TOTAL PARTS

ACCESSORIES

QUAN.	ACCES. NO.	ACCESSORIES	PRICE

TOTAL ACCESSORIES

GRO.-1 ®

Appendix

Section I: DENOMINATE NUMBERS

Denominate numbers are numbers that include units of measurement. The units of measurement are arranged from the largest unit at the left to the smallest unit at the right.

Example: 6 yd 2 ft 4 in

All basic operations of arithmetic can be performed on denominate numbers.

I. EQUIVALENT MEASURES

Measurements that are equal can be expressed in different terms. For example, 12 in = 1 ft. If these equivalents are divided, the answer is 1.

$$\frac{1 \text{ ft}}{12 \text{ in}} = 1 \qquad \frac{12 \text{ in}}{1 \text{ ft}} = 1$$

To express one measurement as another equal measurement, multiply by the equivalent in the form of 1.

To express 6 inches in equivalent foot measurement, multiply 6 inches by 1 in the form of $\dfrac{1 \text{ ft}}{12 \text{ in}}$. In the numerator and denominator, divide by a common factor.

$$6 \text{ in} = \frac{\overset{1}{\cancel{6 \text{ in}}}}{1} \times \frac{1 \text{ ft}}{\underset{2}{\cancel{12 \text{ in}}}} = \frac{1}{2} \text{ ft or } 0.5 \text{ ft}$$

To express 4 feet in equivalent inch measurement, multiply 4 feet by 1 in the form of $\dfrac{12 \text{ in}}{1 \text{ ft}}$

$$4 \text{ ft} = \frac{\overset{4}{\cancel{4 \text{ ft}}}}{} \times \frac{12 \text{ in}}{\underset{1}{\cancel{1 \text{ ft}}}} = \frac{48 \text{ in}}{1} = 48 \text{ in}$$

Per means division, as with a fraction bar. For example, 50 miles per hour can be written $\dfrac{50 \text{ miles}}{1 \text{ hour}}$.

II. BASIC OPERATIONS

A. Addition

Example: 2 yd 1 ft 5 in + 1 ft 8 in + 5 yd 2 ft

1. Write the denomiate numbers in a column with like units in the same column.

2. Add the denominate numbers in each column.

3. Express the answer using the largest possible units.

	2 yd	1 ft	5 in
		1 ft	8 in
+	5 yd	2 ft	
	7 yd	4 ft	13 in

7 yd			=	7 yd		
	4 ft		=	1 yd	1 ft	
		13 in	= +		1 ft	1 in
7 yd	4 ft	13 in	=	8 yd	2 ft	1 in

B. Subtraction

Example: 4 yd 3 ft 5 in − 2 yd 1 ft 7 in

1. Write the denominate numbers in columns with like units in the same column.

	4 yd	3 ft	5 in
−	2 yd	1 ft	7 in

2. Starting at the right, examine each column to compare the numbers. If the bottom number is larger, exchange one unit from the column at the left for its equivalent. Combine like units.

7 in is larger than 5 in

3 ft = 2 ft 12 in

12 in + 5 in = 17 in

3. Subtract the denominate numbers.

	4 yd	2 ft	17 in
−	2 yd	1 ft	7 in
	2 yd	1 ft	10 in

4. Express the answer using the largest possible units.

	2 yd	1 ft	10 in

C. Multiplication

—By a constant

Example: 1 hr 24 min × 3

1. Multiply the denominate number by the constant.

$$
\begin{array}{rr}
\text{1 hr} & \text{24 min} \\
& \times\ 3 \\
\hline
\text{3 hr} & \text{72 min}
\end{array}
$$

2. Express the answer using the largest possible units.

3 hr

$$
\begin{array}{rr}
 & \text{72 min} \\
\hline
\text{3 hr} & \text{72 min}
\end{array}
$$

= 3 hr
= 1 hr   12 min
= 4 hr   12 min

—By a denominate number expressing linear measurement

Example: 9 ft 6 in × 10 ft

1. Express all denominate numbers in the same unit.

$$9 \text{ ft } 6 \text{ in} = 9\frac{1}{2} \text{ ft}$$

2. Multiply the denominate numbers. (This includes the units of measure, such as ft × ft = sq ft.)

$$9\frac{1}{2} \text{ ft} \times 10 \text{ ft} =$$

$$\frac{19}{2} \text{ ft} \times 10 \text{ ft} =$$

95 sq ft

—By a denominate number expressing square measurement

Example: 3 ft × 6 sq ft

1. Multiply the denominate numbers. (This includes the units of measure, such as ft × ft = sq ft and sq ft × ft = cu ft.)

$$3 \text{ ft} \times 6 \text{ sq ft} = 18 \text{ cu ft}$$

—By a denominate number expressing rate

Example: 50 miles per hour × 3 hours

1. Express the rate as a fraction using the fraction bar for *per.*

$$\frac{50 \text{ miles}}{1 \text{ hour}} \times \frac{3 \text{ hours}}{1}$$

2. Divide the numerator and denominator by any common factors, including units of measure.

$$\frac{50 \text{ miles}}{\underset{1}{\cancel{1 \text{ hour}}}} \times \frac{\overset{3}{\cancel{3 \text{ hours}}}}{1}$$

3. Multiply numerators.
    Multiply denominators.

$$\frac{150 \text{ miles}}{1} =$$

4. Express the answer in the remaining unit. 150 miles

D. Division

—By a constant

Example: 8 gal 3 qt ÷ 5

1. Express all denominate numbers
    in the same unit.

    8 gal 3 qt = 35

2. Divide the denominate number
    by the constant.

    35 qt ÷ 5 = 7 qt

3. Express the answer using the
    largest possible units.

    7 qt = 1 gal 3 qt

—By a denominate number expressing linear measurement

Example: 11 ft 4 in ÷ 8 in

1. Express all denominate numbers
    in the same unit.

    11 ft 4 in = 136 in

2. Divide the denominate numbers
    by a common factor. (This includes
    the units of measure, such as inches
    ÷ inches = 1.)

    136 in ÷ 8 in =
    $$\frac{\overset{17}{\cancel{136 \text{ in}}}}{\underset{1}{\cancel{8 \text{ in}}}} = \frac{17}{1} = 17$$

—By a linear measure with a square measurement as the dividend

Example: 20 sq ft ÷ 4 ft

1. Divide the denominate numbers.

    (This includes the units of measure,
    such as sq ft ÷ ft = ft.)

    20 sq ft ÷ 4 ft
    $$\frac{\overset{5 \text{ ft}}{\cancel{20 \text{ sq ft}}}}{\cancel{4 \text{ ft}}} = \frac{5 \text{ ft}}{1}$$

2. Express the answer in the remaining unit. 5 ft

—By denominate numbers used to find rate

Example: 200 mi ÷ 10 gal

1. Divide the denominate numbers.

$$\frac{\frac{200 \text{ mi}}{10 \text{ gal}}}{} = \frac{20 \text{ mi}}{1 \text{ gal}}$$

2. Express the units with the fraction bar meaning *per*.

$$\frac{20 \text{ mi}}{1 \text{ gal}} = 20 \text{ miles per gallon}$$

Note: Alternate methods of performing the basic operations will produce the same results. The choice of method is determined by the individual.

Section II: TABLES

TABLE I EQUIVALENT ENGLISH AND METRIC UNITS OF MEASURE

Linear Measure								
Unit	Inches to millimeters	Millimeters to inches	Feet to meters	Meters to feet	Yards to meters	Meters to yards	Miles to kilometers	Kilometers to miles
1	25.40	0.03937	0.3048	3.281	0.9144	1.094	1.609	0.6214
2	50.80	0.07874	0.6096	6.562	1.829	2.187	3.219	1.243
3	76.20	0.1181	0.9144	9.842	2.734	3.281	4.828	1.864
4	101.60	0.1575	1.219	13.12	3.658	4.374	6.437	2.485
5	127.00	0.1968	1.524	16.40	4.572	5.468	8.047	3.107
6	152.40	0.2362	1.829	19.68	5.486	6.562	9.656	3.728
7	177.80	0.2756	2.134	22.97	6.401	7.655	11.27	4.350
8	203.20	0.3150	2.438	26.25	7.315	8.749	12.87	4.971
9	228.60	0.3543	2.743	29.53	8.230	9.842	14.48	5.592
Example 1 in. = 25.40 mm 1 m = 3.281 ft. 1 km = 0.6214 mi.								

Surface Measure									
Unit	Square inches to square centimeters	Square centimeters to square inches	Square feet to square meters	Square meters to square feet	Square yards to square meters	Square meters to square yards	Acres to hectares	Hectares to acres	Square miles to square kilometers	Square kilometers to square miles
1	6.452	0.1550	0.0929	10.76	0.8361	1.196	0.4047	2.471	2.59	0.3861
2	12.90	0.31	0.1859	21.53	1.672	2.392	0.8094	4.942	5.18	0.7722
3	19.356	0.465	0.2787	32.29	2.508	3.588	1.214	7.413	7.77	1.158
4	25.81	0.62	0.3716	43.06	3.345	4.784	1.619	9.884	10.36	1.544
5	32.26	0.775	0.4645	53.82	4.181	5.98	2.023	12.355	12.95	1.931
6	38.71	0.93	0.5574	64.58	5.017	7.176	2.428	14.826	15.54	2.317
7	45.16	1.085	0.6503	75.35	5.853	8.372	2.833	17.297	18.13	2.703
8	51.61	1.24	0.7432	86.11	6.689	9.568	3.237	19.768	20.72	3.089
9	58.08	1.395	0.8361	96.87	7.525	10.764	3.642	22.239	23.31	3.475
Example 1 sq. in. = 6.452 cm^2 1 m^2 = 1.196 sq. yd. 1 sq. mi. = 2.59 km^2									

Cubic Measure								
Unit	Cubic inches to cubic centimeters	Cubic centimeters to cubic inches	Cubic feet to cubic meters	Cubic meters to cubic feet	Cubic yards to cubic meters	Cubic meters to cubic yards	Gallons to cubic feet	Cubic feet to gallons
1	16.39	0.06102	0.02832	35.31	0.7646	1.308	0.1337	7.481
2	32.77	0.1220	0.05663	70.63	1.529	2.616	0.2674	14.96
3	49.16	0.1831	0.04895	105.9	2.294	3.924	0.4010	22.44
4	65.55	0.2441	0.1133	141.3	3.058	5.232	0.5347	29.92
5	81.94	0.3051	0.1416	176.6	3.823	6.540	0.6684	37.40
6	98.32	0.3661	0.1699	211.9	4.587	7.848	0.8021	44.88
7	114.7	0.4272	0.1982	247.2	5.352	9.156	0.9358	52.36
8	131.1	0.4882	0.2265	282.5	6.116	10.46	1.069	59.84
9	147.5	0.5492	0.2549	371.8	6.881	11.77	1.203	67.32
Example 1 cm^3 = 0.06102 cu. in. 1 gal. = 0.1337 cu. ft.								

Unit	Liquid ounces to cubic centimeters	Cubic centimeters to liquid ounces	Pints to liters	Liters to pints	Quarts to liters	Liters to quarts	Gallons to liters	Liters to gallons	Bushels to hectoliters	Hectoliters to bushels
					Volume or Capacity Measure					
1	29.57	0.03381	0.4732	2.113	0.9463	1.057	3.785	0.2642	0.3524	2.838
2	59.15	0.06763	0.9463	4.227	1.893	2.113	7.571	0.5284	0.7048	5.676
3	88.72	0.1014	1.420	6.340	2.839	3.785	11.36	0.7925	1.057	8.513
4	118.3	0.1353	1.893	8.454	3.170	4.227	15.14	1.057	1.410	11.35
5	147.9	0.1691	2.366	10.57	4.732	5.284	18.93	1.321	1.762	14.19
6	177.4	0.2029	2.839	12.68	5.678	6.340	22.71	1.585	2.114	17.03
7	207.0	0.2367	3.312	14.79	6.624	7.397	26.50	1.849	2.467	19.86
8	236.6	0.2705	3.785	16.91	7.571	8.454	30.28	2.113	2.819	22.70
9	266.2	0.3043	4.259	19.02	8.517	9.510	34.07	2.378	3.171	25.54
Example 1 L = 2.113 pt. 1 gal. = 3.785 L										

TABLE II   DECIMAL EQUIVALENTS

Fraction	Decimal Equivalent Customary (in.)	Metric (mm)	Fraction	Decimal Equivalent Customary (in.)	Metric (mm)
1/64—	0.015625	0.3969	33/64—	0.515625	13.0969
1/32———	0.03125	0.7938	17/32———	0.53125	13.4938
3/64—	0.046875	1.1906	35/64—	0.546875	13.8906
1/16———————	0.0625	1.5875	9/16———————	0.5625	14.2875
5/64—	0.078125	1.9844	37/64—	0.578125	14.6844
3/32———	0.09375	2.3813	19/32———	0.59375	15.0813
7/64—	0.109375	2.7781	39/64—	0.609375	15.4781
1/8———————	0.1250	3.1750	5/8———————	0.6250	15.8750
9/64—	0.140625	3.5719	41/64—	0.640625	16.2719
5/32———	0.15625	3.9688	21/32———	0.65625	16.6688
11/64—	0.171875	4.3656	43/64—	0.671875	17.0656
3/16———————	0.1875	4.7625	11/16———————	0.6875	17.4625
13/64—	0.203125	5.1594	45/64—	0.703125	17.8594
7/32———	0.21875	5.5563	23/32———	0.71875	18.2563
15/64—	0.234375	5.9531	47/64—	0.734375	18.6531
1/4———————	0.250	6.3500	3/4———————	0.750	19.0500
17/64—	0.265625	6.7469	49/64—	0.765625	19.4469
9/32———	0.28125	7.1438	25/32———	0.78125	19.8438
19/64—	0.296875	7.5406	51/64—	0.796875	20.2406
5/16———————	0.3125	7.9375	13/16———————	0.8125	20.6375
21/64—	0.328125	8.3384	53/64—	0.828125	21.0344
11/32———	0.34375	8.7313	27/32———	0.84375	21.4313
23/64—	0.359375	9.1281	55/64—	0.859375	21.8281
3/8———————	0.3750	9.5250	7/8———————	0.8750	22.2250
25/64—	0.390625	9.9219	57/64—	0.890625	22.6219
13/32———	0.40625	10.3188	29/32———	0.90625	23.0188
27/64—	0.421875	10.7156	59/64—	0.921875	23.4156
7/16———————	0.4375	11.1125	15/16———————	0.9375	23.8125
29/64—	0.453125	11.5094	61/64—	0.953125	24.2094
15/32———	0.46875	11.9063	31/32———	0.96875	24.6063
31/64—	0.484375	12.3031	63/64—	0.984375	25.0031
1/2———————	0.500	12.7000	1———————	1.000000	25.4000

TABLE III   CIRCUMFERENCES AND AREAS (0.2 to 9.8; 10 to 99)

Diameter	Circum.	Area	Diameter	Circum.	Area
0.2	0.628	0.0314	31	97.39	754.8
0.4	1.26	0.1256	32	100.5	804.2
0.6	1.88	0.2827	33	103.7	855.3
0.8	2.51	0.5026	34	106.8	907.9
1	3.14	0.7854	35	110	962.1
1.2	3.77	1.131	36	113.1	1,017.9
1.4	4.39	1.539	37	116.2	1,075.2
1.6	5.02	2.011	38	119.4	1,134.1
1.8	5.65	2.545	39	122.5	1,194.6
2	6.28	3.142	40	125.7	1,256.6
2.2	6.91	3.801	41	128.8	1,320.3
2.4	7.53	4.524	42	131.9	1,385.4
2.6	8.16	5.309	43	135.1	1,452.2
2.8	8.79	6.158	44	138.2	1,520.5
3	9.42	7.069	45	141.4	1,590.4
3.2	10.05	7.548	46	144.5	1,661.9
3.4	10.68	8.553	47	147.7	1,734.9
3.6	11.3	10.18	48	150.8	1,809.6
3.8	11.93	11.34	49	153.9	1,885.7
4	12.57	12.57	50	157.1	1,963.5
4.2	13.19	13.85	51	160.2	2,042.8
4.4	13.82	15.21	52	163.4	2,123.7
4.6	14.45	16.62	53	166.5	2,206.2
4.8	15.08	18.1	54	169.6	2,290.2
5	15.7	19.63	55	172.8	2,375.8
5.2	16.33	21.24	56	175.9	2,463
5.4	16.96	22.9	57	179.1	2,551.8
5.6	17.59	24.63	58	182.2	2,642.1
5.8	18.22	26.42	59	185.4	2,734
6	18.84	28.27	60	188.5	2,827.4
6.2	19.47	30.19	61	191.6	2,922.5
6.4	20.1	32.17	62	194.8	3,019.1
6.6	20.73	34.21	63	197.9	3,117.3
6.8	21.36	36.32	64	201.1	3,217
7	21.99	38.48	65	204.2	3,318.3
7.2	22.61	40.72	66	207.3	3,421.2
7.4	23.24	43.01	67	210.5	3,525.7
7.6	23.87	45.36	68	213.6	3,631.7
7.8	24.5	47.78	69	216.8	3,739.3
8	25.13	50.27	70	219.9	3,848.5
8.2	25.76	52.81	71	223.1	3,959.2
8.4	26.38	55.42	72	226.2	4,071.5
8.6	27.01	58.09	73	229.3	4,185.4
8.8	27.64	60.82	74	232.5	4,300.8
9	28.27	63.62	75	235.6	4,417.9
9.2	28.9	66.48	76	238.8	4,536.5
9.4	29.53	69.4	77	241.9	4,656.6
9.6	30.15	72.38	78	245	4,778.4
9.8	30.78	75.43	79	248.2	4,901.7
10	31.41	78.54	80	251.3	5,026.6
11	34.55	95.03	81	254.5	5,153
12	37.69	113	82	257.6	5,281
13	40.84	132.7	83	260.8	5,410.6
14	43.98	153.9	84	263.9	5,541.8
15	47.12	176.7	85	267.0	5,674.5
16	50.26	201	86	270.2	5,808.8
17	53.4	226.9	87	273.3	5,944.7
18	56.54	254.4	88	276.5	6,082.1
19	59.69	283.5	89	279.6	6,221.2
20	62.83	314.1	90	282.7	6,361.7
21	65.97	346.3	91	285.9	6,503.9
22	69.11	380.1	92	289.0	6,647.6
23	72.25	415.4	93	292.2	6,792.9
24	75.39	452.3	94	295.2	6,939.8
25	78.54	490.8	95	298.5	7,088.2
26	81.68	530.9	96	301.6	7,238.2
27	84.82	572.5	97	304.7	7,389.8
28	87.96	615.7	98	307.9	7,543.0
29	91.1	660.5	99	311.9	7,697.7
30	94.24	706.8			

TABLE IV   DRILL SIZES (Letters Z–A, Numbers 1–80)

Letter Size	Decimal Value	Number Size	Decimal Value	Number Size	Decimal Value	Number Size	Decimal Value
Z	0.413	1	0.228	28	0.1405	55	0.052
Y	0.404	2	0.221	29	0.136	56	0.0465
X	0.397	3	0.213	30	0.1285	57	0.043
W	0.386	4	0.209	31	0.12	58	0.042
V	0.377	5	0.2055	32	0.116	59	0.041
U	0.368	6	0.204	33	0.113	60	0.04
T	0.358	7	0.201	34	0.111	61	0.039
S	0.348	8	0.199	35	0.11	62	0.038
R	0.339	9	0.196	36	0.1065	63	0.037
Q	0.332	10	0.1935	37	0.104	64	0.036
P	0.323	11	0.191	38	0.1015	65	0.035
O	0.316	12	0.189	39	0.0995	66	0.033
N	0.302	13	0.185	40	0.098	67	0.032
M	0.295	14	0.182	41	0.096	68	0.031
L	0.290	15	0.18	42	0.0935	69	0.0292
K	0.281	16	0.177	43	0.089	70	0.028
J	0.277	17	0.173	44	0.086	71	0.026
I	0.272	18	0.1695	45	0.082	72	0.025
H	0.266	19	0.166	46	0.081	73	0.024
G	0.261	20	0.161	47	0.0785	74	0.0225
F	0.257	21	0.159	48	0.076	75	0.021
E	0.25	22	0.157	49	0.073	76	0.02
D	0.246	23	0.154	50	0.07	77	0.018
C	0.242	24	0.152	51	0.067	78	0.016
B	0.238	25	0.1495	52	0.0635	79	0.0145
A	0.234	26	0.147	53	0.0595	80	0.0135
		27	0.144	54	0.055		

Glossary

Air bags — Used in suspension systems of trucks and buses. The bags compress air to absorb the shock. Also a safety restraint that pops out of the dash or the steering wheel in an accident.

Alignment — An adjustment to the front and/or rear end of a car that prevents abnormal tire wear and allows the car to be steered easily.

Alternator — An alternating current generator that changes mechanical energy into electrical energy.

Ammeter — An instrument for measuring electric current in amperes.

Ampere — A measure of electrical current.

Ampere-hour — Term used to indicate the capacity of a battery. Product of discharge rate (in amperes) and the time needed (in hours) to discharge a battery.

Antifreeze — Material added to water that lowers water's freezing point and raises its boiling point. Usually is ethylene glycol.

Armature — Rotating part of electrical generator or starter. Coil of wire in electric motor that breaks an electric field.

Axle — A pin or shaft to which a wheel or a pair of wheels is mounted.

Axle ratio — Refers to the number of times the speed is reduced to the ring gear and pinion. Compares the speed of the driveshaft to the speed of the rear axle shaft.

Babbitt material — Soft surface on connecting rods and main bearings in the engine.

BDC — Abbreviation for bottom dead center. The lower limit of the piston motion.

Bearing — A part in which a journal or pin or shaft turns or slides.

Bearing preload — To adjust the pressure of the bearing by a slight overtightening of the bearing adjustment.

Bore — The diameter of a hole, such as a cylinder of an engine. Also, the process of enlarging or refinishing a hole.

Brake horsepower (bhp) — Power delivered to driving wheels.

Braking distance — Distance measured from the time the brakes are applied until the time the car comes to a full stop.

Breaker points — A device or assembly used to direct electrical current within a distributor on older cars.

British thermal units (Btu) — The measurement of the amount of heat needed to raise the temperature of one pound of water one degree Fahrenheit.

Bushing — A removable, soft metal lining placed in a bore to serve as a guide or to limit the size of the opening.

Candlepower — A measurement of the strength of light.

Capacity — The amount of electricity that a battery can deliver. Also refers to the total volume an object can store.

Carbon monoxide (CO) — The colorless, tasteless, odorless but very poisonous gas that forms when gasoline is burned incompletely.

Chassis — Framework of an automobile that includes everything except the car body.

CID — Abbreviation for cubic inch displacement. Number of cubic units (volume) displaced as the piston moves from BDC to TDC. Same as piston displacement.

Circular mil — Unit of area used for the cross-sectional area of electrical wire.

Clearance — Used in connection with parts having circular sections such as pistons, piston pins, and cylinders. Refers to the difference between the two diameters of the parts fitted together.

Clevis pin — A smooth pin with a head on one end and a hole on the other end. Used to connect linkage such as the emergency brake, clutch, or transmission.

Coil — Device used to increase battery voltage to a level that is high enough to jump the spark plug gap.

Coil spring — Length of steel rod wound into a coil. In the suspension system, the springs expand or compress to absorb the shock and support the vehicle.

Compression ratio — Comparison of the amount of space when the piston is at BDC and the amount of space when the piston is at TDC.

Compression stroke — The second step in a four-stroke cycle engine. The valves are closed, and the rising piston compresses the air-fuel mixture.

Computer — An electronic device that receives input signals from many engine sensors and, using this information, readjusts the ignition and fuel systems many times a second.

Condenser — In an electrical system, a device that temporarily collects and stores electrical current for later discharge. In an air-conditioning system, a device that changes refrigerant gas into a liquid.

Connecting rod — The rod that connects the piston to the crankshaft.

Countershaft — A shaft that receives motion from a mainshaft and transmits it to a working part.

Crankshaft — The main rotating member, or shaft, of the engine, which changes the up and down motion of the pistons into rotary motion. Sometimes called the crank.

Crossmember — The strips of steel that attach the side members of an automobile frame to each other like the rungs of a ladder.

Cycle — A series of repeated events. In a four-stroke cycle engine, it consists of the intake, compression, power, and exhaust strokes.

Cylinder — The chamber in which the piston travels.

Cylinder block — The main part of the engine in or on which other engine parts are attached.

Cylinder head — The part of the engine that encloses the tops of the cylinders.

Dead center — Extreme high or low position in the crankshaft throw when the piston is not moving.

Differential — The part of the rear axle assembly that allows one wheel to turn at a different speed than the other while transmitting power from the driveshaft to the rear axles.

Direct drive — Refers to when the engine and the driveshaft are completely in mesh, and the crankshaft and driveshaft turn at the same speed.

Distributor — A device that directs the current from the coil to the spark plugs of the engine in the proper firing order.

Driveshaft — Revolving shaft that transmits motion. In automotive use, shaft that connects the transmission to the rear axle assembly. Sometimes referred to as propeller shaft.

Dwell — The number of degrees the distributor shaft rotates while current is flowing in the coil primary winding. On older cars, it is the number of degrees the distributor shaft rotates while the points remain closed.

Efficiency — The comparison of output to input.

EGR valve — The exhaust gas recirculation valve is an emission control device that allows a small amount of exhaust gas back into the intake manifold to lower the oxides of nitrogen (NOX).

Ethylene glycol — A thick, liquid alcohol that when mixed with water lowers the freezing point and raises the boiling point of the coolant. Sometimes mistakenly called permanent antifreeze.

Exhaust emissions — The unburned hydrocarbons that escape through the tailpipe of a car.

Exhaust manifold — The part of the engine that collects the exhaust gases from the cylinders and carries the gases to the exhaust pipe.

Exhaust stroke — The last step in a four-stroke cycle engine. The exhaust valve is open, and the rising piston pushes gas out of the cylinder.

Feeler gauge — A thin blade of precise thickness that is used to measure the clearance between two parts.

Flat rate — Method of charging for automotive repairs based on the standard time required and a set amount per job. The actual time taken to do the job is not considered.

Flow test — The test used to find out if a radiator is clogged.

Flywheel — A heavy metal wheel attached to the crankshaft that smooths out the power impulses of the engine.

Foot-pound — A measurement of work. The work done in moving one pound a distance of one foot.

Gasket — A flat piece of fiber, cork, or metal placed between two surfaces to act as a tight seal.

Gauge — An instrument with a graduated scale or dial used to measure object or show quantities.

Gear ratio — The comparison of numbers of teeth on gears or the diameters on pulleys. Also refers to a comparison of the speeds of gears.

Generator — A device that changes mechanical energy into electrical energy.

High-compression head — An engine with less than the standard space above the piston. The mixture is compressed more tightly, and the engine has greater power.

Hone — A device with fine grinding stones powered by an electric drill that is used to increase the diameter of a cylinder a few thousandths of an inch.

Horsepower (hp) — A measurement of the rate at which work is done. The energy required to lift 550 pounds a distance of one foot in one second of time. One unit of horsepower equals 33,000 foot-pounds of work per minute.

Hub — The central part of a rotating object such as a wheel.

Hydrocarbon (HC) — An emission resulting from incomplete combustion. Hydrocarbons are emitted when the spark plugs fail to burn the mixture in the cylinder completely.

Idler gear — The gear between the input and output shafts of a transmission that changes the rotation of the output shaft but does not affect the speed.

Ignition coil — The part of the electrical system that produces the spark that ignites the mixture in the cylinders at exactly the right time.

Indicated horsepower (ihp) — The horsepower delivered to the piston by the burning gas.

Intake manifold — The part of the engine that carries the air-fuel mixture from the carburetor to the cylinders.

Intake stroke — The first step in the four-stroke cycle engine. The inlet valves are open, and the pistons draw the fuel and air mixture into the cylinder.

Interference fit — When the hole is smaller than the pin and the pin must be pressed in with a press.

Jack — A device used to raise the car so the wheels can be removed or repairs can be made.

Journal — The highly machined surface of a rotating shaft that turns in a bearing.

Leaf spring — An automotive spring that is made with one or more strips of spring steel.

Lockwasher — A special washer placed under the head of a bolt or nut to prevent the bolt or nut from working loose.

Low-compression head — An engine with more than the standard space above the piston. The mixture is compressed loosely, and the engine has less power.

Main (journal) bearings — The bearings that hold the crankshaft to the engine block.

Medium-compression head — An engine with the standard space above the piston. The mixture is compressed normally, and the engine has standard power.

Micrometer — A precision measuring instrument for internal and external measurements to a thousandth inch. Also available in metric measurement.

Mill — One-thousandth (1/1000) dollar or one-tenth (1/10) cent.

Misfire — Failure of spark plug to ignite a charge at the proper time.

Module, Electronic ignition — A solid-state electronic device that controls the coil primary circuit.

NC — Abbreviation for National Coarse. A thread series of a few coarse threads per inch.

NF — Abbreviation for National Fine. A thread series of many fine threads per inch.

Nitrogen oxide (NOX) — A by-product of combustion that occurs when the combustion chamber is at a high temperature and under a heavy load. Also called oxides of nitrogen.

Odometer — The part of the speedometer head that records the miles traveled.

Ohm — The measurement of electrical resistance.

Ohm's Law — A principle of the relationship between current, voltage, and resistance. The formula is written $I = \dfrac{E}{R}$.

Overhaul — To repair a vehicle thoroughly.

Oversize — Used in connection with parts having circular sections, such as pistons, piston pins, and cylinders. Refers to an increase in the diameter of the part from the original or standard size.

PCV valve — Positive crankcase ventilation valve is an emission-control device that returns crankcase emissions to the intake manifold to be burned in the engine.

Pinion gear — The smaller of two meshing gears. The gear that drives the ring gear in the rear axle assembly.

Piston — The cylinder-shaped part connected to the crankshaft by a connecting rod. The force of explosion in the cylinder exerts a force on the piston, and the crankshaft moves up and down

Piston displacement — The volume (number of cubic inches) displaced as the piston moves from BDC to TDC.

Piston pin — The pin that connects the piston to the connecting rod. Sometimes called the wrist pin.

Pitch — The distance between the threads on a bolt or screw.

Pop rivet — A rivet used to secure metal when it is not possible to get to the back side.

Radial-ply tire — A tire in which the layers are laid perpendicular to the rim and a belt is placed around the circumference.

Rebate — A refund from a manufacturer for purchasing their product.

Refrigerant — A chemical used in the air-conditioning system of a vehicle that absorbs and gives heat as it changes from liquid to gas state and gas to liquid state.

Resistance — The opposition of a substance to the passage of electrical current.

Revolution — Circular movement of an object about a center or axis.

Ring gear — The largest gear in the rear axle assembly.

Rotor — In the ignition, the part attached to the distributor shaft that directs current to the distributor cap terminals. In disc brakes, the steel disc attached to the wheels.

Rpm — Abbreviation for revolutions per minute. Measure of the number of times a shaft turns per minute.

SAE — Abbreviation for Society of Automotive Engineers.

SAE horsepower (SAE hp) — Method of estimating horsepower, usually used for taxation purposes.

Scanner — An electronic diagnostic instrument used on cars equipped with computers.

Severe service — High speeds at high temperatures and loads.

Shaft — A cylinder-shaped bar that transmits power or motion by rotation.

Socket — Interchangeable end for a wrench that covers the nut and applies pressure on all corners.

Spark plug — The automotive part that fits into the cylinder head of an engine. The two electrodes are separated by an air gap across which the current from the ignition system jumps and forms the spark for combustion.

Specific gravity — The weight of a substance compared to an equal amount of water.

Speedometer — An instrument used to measure and indicate the speed of a vehicle such as in miles per hour or kilometers per hour.

Speed wrench — A wrench used with sockets that has an offset crank that allows the handle to be turned rapidly.

Splines — Slots or grooves cut into a shaft. A shaft with grooves cut lengthwise allowing separate sliding motion but preventing circular motion without turning another part.

Spring — Flexible members that support the vehicle weight. The four types of springs are coil, leaf, torsion bar, and air bags.

Sprocket — A toothed wheel shaped so it will interlock with a chain.

Stall (bay) — A work space in a garage that is large enough for one car.

Standard size — The original size or the size based on manufacturers' specifications.

Standard taper pins — Pins with a taper equal to $1/4$ inch per foot.

Starter commutator — The part of the armature on which the brushes ride.

Stock — A piece of metal. Also, the same as the standard or factory item.

Stroke — In the engine, the distance the piston moves from TDC to BDC or the distance the piston moves in one-half a crankshaft revolution. Mathematically equal to twice the length of the crankshaft throw.

Tap — A hardened steel tool for cutting internal threads in a hole.

Taper — The difference in the diameters between two ends of a pin.

TDC — Abbreviation for top dead center. The upper limit of the piston motion.

Technician — A specialist in the technical details of an occupation. A better description of the automobile mechanic who repairs the modern automobile.

Terminals — Points, such as battery terminals or posts, where electrical connections are made.

Thermostat — A device that operates on or regulates temperature changes. In the cooling system, it restricts circulation until the engine is at normal operating temperature.

Throw — The distance from the center of the crankshaft main bearing journal to the center of the connecting rod journal.

Toe-in — The adjustment of the front wheels of a car so that the front of the wheels are closer together than the back of the wheels.

Torque — A turning or twisting force that produces rotation.

Torsion bar — A part of the automobile suspension system. One end of the bar is fixed to the frame, and the other is twisted and connected to the lower control arm acting as a spring.

Total gear reduction — Comparison of the speed of the crankshaft to the speed of the rear axle shaft.

Transmission — The part of the power train that includes the speed-changed gears by which power is transferred from the engine to the axle.

Transmission gear ratio — Refers to the number of times speed is reduced by the transmission. Comparison of the speed of the crankshaft to the speed of the driveshaft.

Tubing — A pipe made of metal or synthetic rubber that is used to move liquids or a vacuum.

Tune-up — The process of carefully and accurately adjusting automotive parts and systems in order to get maximum engine performance.

Turbocharger — A turbine driven by exhaust gases that forces air under pressure into the air intake of an engine, thus boosting the engine's power output.

Turning radius — The radius of the circle within which a car can be turned around. The distance from the center of the rear axle to the pivot point.

TVRS wire — Spark plug wiring that will not cause television or radio interference.

Undersize — Refers to a decrease in the diameter of the part from the original or standard size. In bearings, only the inside diameter is smaller.

Universal joint — The connection that allows a driveshaft to drive at an angle. It can also be found in some steering shafts.

Vacuum — An absence of air or any other material. Less than atmospheric pressure.

Valve — A device for opening or closing a passageway. In the engine, the intake valve allows the mixture into the cylinder, and the exhaust valve lets it out into the manifold.

Viscosity — A liquid's resistance to flow. Used especially with weights of oil.

Volt — Measure of electrical force. Force needed to cause a current of one ampere to flow through a resistance of one ohm.

Washer — A thin ring used between parts to ensure tightness.

Weather strip — A strip of material, usually synthetic rubber, used around doors and trunk lids to keep out dust and water.

Wrist pin — A common name for a piston pin.

ANSWERS TO ODD-NUMBERED PROBLEMS

SECTION 1   WHOLE NUMBERS

UNIT 1   ADDITION OF WHOLE NUMBERS

1. 61
3. 383
5. 1144
7. 30 feet

9. 187 clamps
11. 50 gallons
13. 24 amperes
15. 482 kilometers

UNIT 2   SUBTRACTION OF WHOLE NUMBERS

1. 464
3. 22
5. 276
7. 4764
9. 311 quarts

11. 39 gallons
13. $223
15. 2 meters
17. 83 plugs
19. 258 miles

UNIT 3   MULTIPLICATION OF WHOLE NUMBERS

1. 171
3. 292
5. 17,361
7. 88,452
9. 432 miles

11. 3,843 feet
13. 5,373 miles
15. 504 clips
17. 648 rivets

UNIT 4   DIVISION OF WHOLE NUMBERS

1. 56
3. 9.5
5. 85
7. 15
9. 6 cars

11. 12 gallons
13. 21 mpg
15. 660 revolutions
17. 3 weeks
19. 5 liters per car

SECTION 2   DECIMAL FRACTIONS

UNIT 5   FRACTIONAL EQUIVALENTS

1. a. $\frac{7}{8}$ inch
   b. $\frac{9}{16}$ inch
   c. $\frac{15}{64}$ inch
   d. $\frac{5}{8}$ inch
   e. $\frac{13}{16}$ inch
   f. $\frac{7}{16}$ inch
   g. $\frac{15}{64}$ inch
   h. $\frac{3}{4}$ inch
   i. $\frac{1}{32}$ inch
   j. $\frac{3}{32}$ inch
   k. $\frac{11}{16}$ inch
   l. $\frac{15}{16}$ inch
3. $\frac{5}{16}$ inch
5. $\frac{1}{2}$ inch
7. $\frac{5}{8}$ inch
9. $\frac{13}{16}$ inch

UNIT 6   ADDITION OF DECIMAL FRACTIONS

1. 5.249 inches
3. $47.04
5. 102.395 millimeters
7. 14.750 inches
9. 3.880 inches
11. 5.125 inches
13. 6.875 inches
15. 0.675 inch
17. + 0.020 inch
19. 17.46 millimeters
21. 3.395 inches
23. 3.880 inches
25. 0.852 inch
27. 3.693 inches
29. 4.139 inches
31. 17.447 millimeters

UNIT 7   SUBTRACTION OF DECIMAL FRACTIONS

1. 0.250 inch
3. 0.0035 inch
5. 1.93 millimeters
7. 3.766 inches
9. 200.92 millimeters
11. 4.5 inches
13. 2.445 inches
15. 0.15 kilogram per square centimeter
17. 0.002 inch
19. 0.0100 inch
21. 0.0035 inch
23. 4.028 inches
25. 0.3575 inch

UNIT 8   MULTIPLICATION OF DECIMAL FRACTIONS

1. 7
3. 158.02
5. .62546
7. $\frac{13}{16}$ inch nearest
9. $\frac{25}{32}$ inch nearest
11. $\frac{11}{32}$ inch nearest
13. $\frac{5}{8}$ inch nearest
15. $\frac{1}{16}$ inch nearest
17. $\frac{13}{16}$ inch nearest
19. 25.395 millimeters

21. $^{48}\!/_{64}$ inch nearest
23. 12.66 volts
25. $70.83
27. $145.20
29. $344.00
31. $13.04
33. $1,910,595.50
35. $790.79

37. $50,256.25
39. $927.30
41. 7,452 pounds
43. $120.00
45. $63.00
47. 101.20 ounces
49. 682.1 pounds
51. $69.13

UNIT 9 DIVISION OF DECIMAL FRACTIONS

1. $10.80 per hour
3. 72.26 kilometers per hour
5. $11.37 per hour
7. 9.7 hours
9. 6.56 pounds per gallon
11. 6 quarts
13. $0.006 per mile

15. $0.001 per mile
17. 65.04 turns
19. $0.85 per foot
21. 3.721 inches per space
23. $28.58 per payment
25. 4.88 hours

UNIT 10 MICROMETER READING: APPLICATION OF DECIMAL FRACTIONS

1. 8–3–0
3. 2–0–18¾
5. 6–2–16
7. 0–3–2
9. 6"–7"; no; 1–3–12½
11. 3"–4"; yes; 2–2–0
13. 4"–5"; no; 0–2–17
15. 3"–4"; yes; 4–3–24
17. 21–0–21
19. 19–1–17
21. 25–0–40

23. 0.240
25. 0.800
27. 0.599
29. 0.434
31. 0.898
33. 3.245"
35. 2.374"
37. 4.999"
39. 18.95 mm
41. 14.03 mm

SECTION 3   COMMON FRACTIONS

UNIT 11   MULTIPLICATION OF COMMON FRACTIONS

1. $425\frac{1}{4}$ pounds
3. $449\frac{1}{4}$ inches
5. 6 pounds
7. $684\frac{1}{2}$ miles
9. $30\frac{5}{8}$ seconds
11. $\frac{1}{2}$ inch
13. $\frac{3}{8}$ inch
15. $88\frac{3}{4}$ inches

17. 1,015 inches
19. $10\frac{1}{8}$ inches
21. $14\frac{1}{16}$ pounds
23. 8 quarts antifreeze
25. $3\frac{11}{16}$ inches
27. $15\frac{3}{4}$ inches
29. 15 inches
31. $718\frac{1}{10}$ miles

UNIT 12   DIVISION OF COMMON FRACTIONS

1. $7\frac{1}{4}$ inches
3. $4\frac{1}{4}$ inches
5. $2\frac{7}{64}$ inches
7. $\frac{7}{8}$ hour per car
9. 5 cars
11. 20 strokes
13. 32 miles per gallon
15. $12\frac{1}{4}$ hours
17. $12.40 per hour
19. 10 turns

21. $4\frac{1}{8}$ inches
23. 155 nuts
25. $1\frac{7}{8}$ inches
27. 10 pieces
29. 35 bushings
31. 180 shims
33. 7 lengths
35. 30 blanks
37. 100 bolts

UNIT 13   ADDITION OF COMMON FRACTIONS

1. $32\frac{3}{4}$ inches
3. $32\frac{1}{2}$ inches
5. $9\frac{1}{2}$ hours
7. $10\frac{7}{8}$ tons
9. $\frac{11}{32}$ inch
11. $\frac{7}{32}$ inch
13. $1\frac{3}{4}$ inches
15. $1\frac{5}{16}$ inches
17. 6 inches
19. $6\frac{7}{32}$ inches
21. $408\frac{1}{2}$ centimeters

23. $53\frac{51}{64}$ inches
25. $2\frac{3}{16}$ inches
27. $4\frac{3}{4}$ inches
29. $1\frac{5}{8}$ inches
31. $\frac{13}{16}$ inch
33. $23\frac{3}{16}$ inches
35. $38\frac{19}{32}$ inches
37. $50\frac{9}{32}$ inches
39. $2\frac{17}{32}$ inches
41. $254\frac{1}{10}$ centimeters

UNIT 14   SUBTRACTION OF COMMON FRACTIONS

1. $1\frac{7}{16}$ inches
3. $1\frac{5}{16}$ inches
5. $\frac{1}{64}$ inch
7. $\frac{13}{16}$ inch
9. $\frac{7}{8}$ inches
11. $4\frac{5}{8}$ inches
13. $13\frac{5}{8}$ inches
15. $57\frac{1}{2}$ miles
17. $\frac{1}{8}$ inch
19. $\frac{5}{8}$ inch
21. $\frac{25}{32}$ inch
23. $\frac{31}{32}$ inch
25. $1\frac{7}{16}$ inches
27. $1\frac{3}{16}$ inches
29. $1\frac{5}{16}$ inches
31. $1\frac{7}{16}$ inches
33. $\frac{1}{8}$ inch
35. $6\frac{2}{5}$ meters
37. $2\frac{1}{4}$ inches
39. $\frac{7}{8}$ gallon
41. $10\frac{1}{64}$ inches

43. $2\frac{1}{4}$ horsepower
45. $\frac{3}{8}$ inch
47. $\frac{3}{4}$ inch
49. $\frac{15}{16}$ inch
51. $\frac{15}{16}$ inch
53. $\frac{5}{16}$ inch
55. $2\frac{3}{16}$ inches
57. $2\frac{7}{16}$ inches
59. $4\frac{1}{16}$ inches
61. $2\frac{5}{16}$ inches
63. $4\frac{15}{16}$ inches
65. $\frac{5}{8}$ inch
67. $1\frac{3}{16}$ inches
69. $2\frac{1}{8}$ inches
71. $\frac{5}{8}$ inch
73. $2\frac{15}{16}$ inches
75. $2\frac{3}{16}$ inches
77. $4\frac{1}{4}$ inches
79. $2\frac{1}{16}$ inches
81. $1\frac{13}{16}$ inches
83. $1\frac{7}{8}$ inches

UNIT 15   MULTIPLE OPERATIONS OF COMMON AND DECIMAL FRACTIONS

1. $368\frac{1}{2}$ miles per day
3. 4 stalls
5. 3.925 inches
7. 3.530 inches
9. a. 0.075 inch
   b. $\frac{1}{16}$ inch
11. $\frac{5}{16}$ inch nearest
13. $102.95
15. $0.31 per mile
17. 3.53 times filled
19. $257.65
21. $662.43
23. 13 cars
25. 0.8125 inch

27. 0.938 inch
29. 0.750 inch
31. $\frac{55}{64}$ inch nearest
33. 0.0625 inch
35. 0.844 inch
37. 0.766 inch
39. 0.047 inch
41. a. 0.257 inch
    b. F
43. 0.453 inch
45. $\frac{37}{64}$ inch
47. #10
49. 2 times more copper
51. 4 times more copper

SECTION 4   PERCENT AND PERCENTAGE

UNIT 16   SIMPLE PERCENT

1. ¼
3. ¹⁄₁₀
5. ⅗
7. ⅔
9. ¹⁄₂₀
11. 1¼
13. ¹⁹⁄₂₀
15. ¹⁄₅₀
17. 0.67
19. 0.0175
21. 0.125
23. 0.435
25. 0.87
27. 5
29. 0.48
31. 0.17
33. 1.25
35. 0.03
37. 25%
39. 10%
41. 5%

43. 23.2%
45. 25%
47. 21.4% rounded
49. 41.67%
51. 3.2% rounded
53. 21.4%
55. 9%
57. 18.4% rounded
59. 26.3% rounded
61. 40%; −12°F
63. 33⅓%; 0°F
65. 25%; +10°F
67. 20%; +16°F
69. 40%; −12°F
71. 60%; −62°F
73. 33⅓%; 0°F
75. 20%
77. 12.5%
79. 36.6%
81. 23.6%
83. 40%

UNIT 17   SIMPLE PERCENTAGE

1. $11.85
3. $28.60
5. $6.02
7. 50 taillights
9. 10 cans
11. 1.25 pounds
13. 37.80 pounds
15. 204
17. 85

19. 293
21. $48.27
23. 291 rounded
25. $18,864.81
27. 90,000
29. 17%
31. 64
33. 12 cars

UNIT 18   DISCOUNTS

1. $18.81
3. $8.28
5. $5.85
7. $90.97
9. $64.50
11. $1,438.27
13. $63.43

15. $157.04
17. 23.3%
19. 14.9%
21. $658.19
23. $111.49
25. 13.29%

UNIT 19   PROFIT AND LOSS, COMMISSIONS

1. a. yes
   b. 25%
   c. $1,500.00
3. $1,002.33
5. $2,213.75
7. $27.94
9. $1.45
11. $1,831.50

13. $565.59
15. 10%
17. 5,559.12
19. 23%
21. $28.87
23. $4,260.00
25. $88.20

UNIT 20   INTEREST AND TAXES

1. $21.75
3. a. $157.50
   b. $1,157.50
   c. 15.75%

5. 2%
7. 2.95%
9. $49
11. 10.3 years (rounded)

UNIT 21   PERCENT OF ERROR AND AVERAGES

1. 8.7%
3. 0.008 inch
5. 397.75 miles

7. 77.2%
9. 32,973.6 miles

SECTION 5   MEASUREMENT

UNIT 22   ENGLISH LINEAR MEASUREMENTS

1. $114\frac{5}{8}$ inches
3. 2 feet $6\frac{1}{4}$ inches
5. 50 feet $10\frac{1}{2}$ inches
7. 1,760 yards
9. 11,880.00 yards

11. 25.5 inches
13. 134,112.0 feet
15. 4,400 yards
17. 5.5 feet
19. 14 inches

UNIT 23   METRIC MEASUREMENTS

1. 100 centimeters
3. 2,500 millimeters
5. 260 millimeters
7. 1.44 inches
9. 101.6 millimeters
11. 11 threads per centimeter
13. 3.157 inches
15. $\frac{3}{4}$"
17. 8.070"

19. 3.175 millimeters
21. yes
23. 3.15"
25. 3.54"
27. $\frac{3}{8}$"
29. no
31. yes; 1.377" diameter
33. 1.365" or $1\frac{23}{64}$" nearest

UNIT 24   SCALE READING

1. 64
3. 10
5. 4
7. 6
9. $\frac{9}{16}$ inch
11. $2\frac{7}{16}$ inches
13. $1\frac{1}{4}$ inches
15. $2\frac{5}{8}$ inches
17. $3\frac{3}{8}$ inches
19. $4\frac{1}{2}$ inches
21. 13 millimeters
23. 81 millimeters

25. 142 millimeters
27. 2.0 centimeters
29. 8.6 centimeters
31. $1\frac{1}{16}$ inches
33. $1\frac{7}{8}$ inches
35. $1\frac{3}{16}$ inches
37. 42 millimeters
39. 81 millimeters
41. 8 millimeters
43. $4\frac{7}{8}$ inches
45. 50 millimeters

UNIT 25   SCALE READING OF TEST METERS

1. 37 degrees
3. 8000 ohms
5. 1.5 ohms
7. 250 PPM HC; 2% CO
9. 9.5 volts at 200 amps

11. 1800 PPM HC; 2% CO
13. 14°
15. 22.6 volts
17. 13.8 volts

UNIT 26   CIRCULAR MEASUREMENT

Note: All problems are solved using C = (3.1416)(D). Answers may vary if alternate method is used.

1. 9.425 inches
3. 11.781 inches
5. 11 inches

7. 1.751 inches
9. 0.687 inch

UNIT 27   ANGULAR MEASUREMENT

1. 180°
3. $\frac{1}{6}$
5. $\frac{1}{12}$
7. 45°
9. 45°
11. 24°
13. 51° 25' 43"

15. 60°
17. 120°
19. 132°
21. 239°
23. 1$\frac{9}{35}$ inches or 1.257 inches
25. 3$\frac{2}{3}$ inches or 3.665 inches

UNIT 28   AREA AND VOLUME MEASUREMENT

1. 30,000 square feet
3. 10 feet
5. 420 square inches
7. 248 square feet
9. 3,300 square inches
11. $47,005.00

13. 12.4995 pounds
15. 13.1 gallons
17. 7.48 gallons
19. 6,750 pounds
21. 59.69 kg (131.33 lb)
23. 1363 kg

UNIT 29 TIME, SPEED, AND MONEY CALCULATIONS

1. $73.46
3. 24 jobs
5. 5 weeks 1 day 2 hours
7. 0.4 hour
9. 54 miles per hour

11. 62.6 kilometers per hour
13. 88.00 feet per second
15. 59.30 kilometers per hour
17. 52.1 knots
19. 2,266.67 feet per minute

SECTION 6 RATIO AND PROPORTION

UNIT 30 RATIOS

1. 3:1
3. 1.8:1
5. 1.875:1
7. 7 times faster
9. 4:1
11. 2:1
13. 8.6:1
15. 3.25:1
17. 1:1

19. 3.5:1
21. 2.5:1
23. 8.46:1
25. 6.82:1
27. 437.5 rpm
29. 4,000 rpm
31. 3,214.3 miles
33. 20.17:1
35. 2.8:1

UNIT 31 PROPORTIONS

1. 800 revolutions per minute
3. 70 teeth
5. 1385 revolutions per minute
7. 250 revolutions per minute
9. 1560 revolutions per minute
11. 200 revolutions per minute
13. clockwise
15. 1200 revolutions per minute
17. 675 revolutions per minute
19. 2250 revolutions per minute
21. 1482 revolutions per minute
23. 8889 revolutions per minute
25. 12 inches

27. 3.2 centimeters
29. 187.5 inches
31. 30 inches
33. $^{104}/_{1000}$ inches
35. 3.000 inches
37. 1675 revolutions per minute
39. 22 feet per second
41. 36.66 feet per second
43. 22.5 miles per hour
45. a. yes
    b. correct
47. 8 minutes
49. 182.4 miles

SECTION 7 FORMULAS

UNIT 32 FORMULAS FOR CIRCULAR MEASUREMENT

1. a. 37.699 inches
   b. 6 inches
3. a. 20 inches
   b. 62.832
5. 193.21 millimeters
7. a. 26.10 inches
   b. 82 inches

   c. 1.078 extra turns
9. 14.6608 feet further
11. 42 feet 8 inches
13. 0.196 square meter
15. 41.283 square inches
17. 4.03 feet
19. 0.528 inch

UNIT 33 FORMULAS FOR EFFICIENCY

1. 74 percent
3. 77 percent

5. 24 percent

UNIT 34 TEMPERATURE

1. 20° Celsius
3. 2,300.0° Fahrenheit

5. 620.6° Fahrenheit
7. 1,504.4° Celsius

UNIT 35 CYLINDRICAL VOLUME MEASUREMENT

1. 10.2 gallons
3. 100 gallons
5. 235.3 inches
7. 336.740 cubic inches

9. 3.56 inches
11. 121.85 cubic inches
13. gasoline; 537.88 cc larger
15. 272.591 cubic inches

UNIT 36 HORSEPOWER

1. 25.1 horsepower
3. doubles the horsepower

5. 33.8 horsepower

UNIT 37 OHM'S LAW

1. 2.136 ohms
3. 1,333 amps
5. 7.88 amps rounded

7. 4.24 amps
9. 0.088 ohm rounded
11. 136 amps

SECTION 8 GRAPHS

UNIT 38 USE OF GRAPHS

1. 110 feet
3. 176 feet
5. $1,700
7. $1,500
9. $33\frac{1}{3}\%$

11.

PARTS OUTPUT

_____ Year One − − − − Year Two

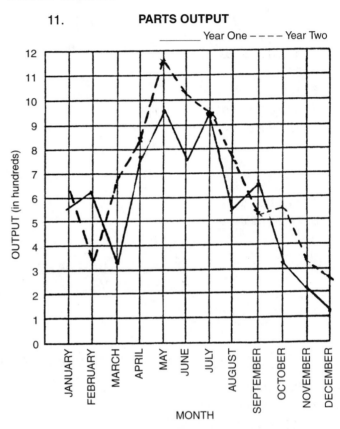

13. 65 horsepower
15. 2,500 revolutions per minute
17. 3,400 revolutions per minute
19. 1,750 revolutions per minute
21. 4.5 seconds
23. 82 pounds
25. 12 pounds approximately

27. 3,375 revolutions per minute
29. 1,040° Fahrenheit approximately
31. 110° Fahrenheit
33. oxides of nitrogen (No_x)
35. decreases
37. approximately 25%

39.

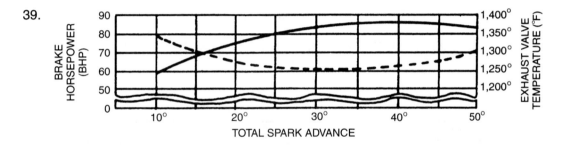

41. increases

43. small effect on brake horsepower

45.

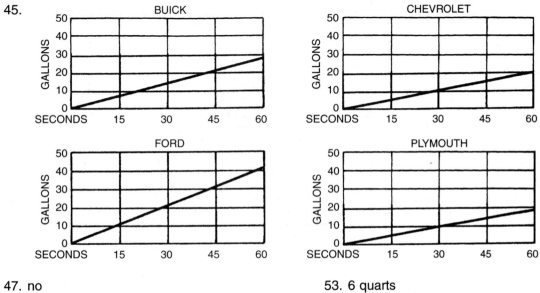

47. no

49. 7 seconds

51. 6 quarts

53. 6 quarts

55. 7 quarts

SECTION 9   INVOICES

UNIT 39   REPAIR ORDERS AND SHOP TICKETS

PROBLEM 1.

1.

\u00A0	MATERIAL USED				
QUAN.	PART NO.	DESCRIPTION	PRICE		
8		SPARK PLUGS	2	98	23 84
1	SET SPARK	PLUG WIRES			55 83
1		DIST. CAP			18 90
1		ROTOR			6 80
1		CARB. KIT			33 15

MIDTOWN GARAGE
212 NORTHERN BLVD.
BROWNVILLE, NEW YORK 13000
TELEPHONE 853-4008

REPAIR ORDER INSTRUCTIONS

TUNE-UP LABOR — 42 00
REBUILD CARBURETOR — 45 00

OUTSIDE REPAIRS

BROUGHT FORWARD

TOTAL PARTS — $138 02

TOTAL LABOR — 87 00
TOTAL PARTS — 138 02
ACCESSORIES
GAS, OIL & GREASE
OUTSIDE REPAIRS
5 % TAX — 6 90
TOTAL AMOUNT — $231 92

QUAN.	ACCES. NO.	ACCESSORIES	PRICE		

TOTAL ACCESSORIES

2. $43.50

3. $13.59

4. $55.43

PROBLEM 3.

1.

MATERIAL USED					
QUAN.	PART NO.	DESCRIPTION	PRICE		
30"	5/8"	HOSE	FT/1.25	3	13
24"	3/4"	HOSE	FT/1.35	2	70
2	5/8"	CLAMPS	.84	1	68
2	3/4"	CLAMPS	.84	1	68
1		THERMOSTAT		5	65
1		CAP.		4	50
1		UPPER HOSE		11	44
1		LOWER HOSE		16	53
4		HOSE CLAMPS	.98	3	92

TOTAL PARTS $51.23

MIDTOWN GARAGE

212 NORTHERN BLVD.
BROWNVILLE, NEW YORK 13000
TELEPHONE 853-4008

OPER. NO.	REPAIR ORDER INSTRUCTIONS	LABOR CHARGE
	REMOVE, CLEAN, REPLACE RADIATOR	39 50
	REVERSE FLUSH BLOCK	20 00
	REPLACE HEATER HOSES	10 00
	REPLACE THERMOSTAT & CAP	N/C

TOTAL LABOR	69 50
TOTAL PARTS	51 23
7% TAX	3 59
TOTAL AMOUNT	124 32

2. $41.70

3. $13.90

4. $30.74

5. $48.29

6. 42.4%